STUDENT SOLUTIONS MANUAL
for
PROBABILITY AND STATISTICS
FOR ENGINEERING AND THE SCIENCES

Fourth Edition

Jay L. Devore
California Polytechnic State University
San Luis Obispo

Elizabeth M. Eltinge
Texas A & M University

Duxbury Press
An Imprint of Wadsworth Publishing Company
I(T)P® An International Thomson Publishing Company

Belmont • Albany • Bonn • Boston • Cincinnati • Detroit • London • Madrid
Melbourne • Mexico City • New York • Paris • San Francisco
Singapore • Tokyo • Toronto • Washington

Printed in the United States of America
2 3 4 5 6 7 8 9 10

For more information, contact Duxbury Press at Wadsworth Publishing Company.

Wadsworth Publishing Company
10 Davis Drive
Belmont, California 94002, USA

International Thomson Publishing
Europe
Berkshire House 168-173
High Holborn
London, WC1V 7AA, England

Thomas Nelson Australia
102 Dodds Street
South Melbourne 3205
Victoria, Australia

Nelson Canada
1120 Birchmount Road
Scarborough, Ontario
Canada M1K 5G4

International Thomson Editores
Campos Eliseos 385, Piso 7
Col. Polanco
11560 México D.F. México

International Thomson Publishing
GmbH
Königswinterer Strasse 418
53227 Bonn, Germany

International Thomson Publishing Asia
221 Henderson Road
#05-10 Henderson Building
Singapore 0315

International Thomson Publishing Japan
Hirakawacho Kyowa Building, 3F
2-2-1 Hirakawacho
Chiyoda-ku, Tokyo 102, Japan

Compositor: Laurel Technical Services

ISBN 0-534-24265-0

TABLE OF CONTENTS

CHAPTER 1

Section 1.1

1 **a** Houston Chronicle, Des Moines Register, Chicago Tribune, Washington Post

b Eaton, Eli Lilly, General Motors, Ford

c Bill Jasper, Kay Reinke, Helen Ford, David Menendez

d KAMU, KCHI, KKIK, KUSF

3 Concrete: All living U.S. Citizens, all mutual funds marketed in the U.S., all books published in 1980.

Hypothetical: All grade point averages for University of California undergraduates during the next academic year. Page lengths for all books published during the next calendar year. Batting averages for all major league players during the next baseball season.

5 One could take a simple random sample of students from all students in the California State University system and ask each student in the sample to report the distance form their hometown to campus. Alternatively, the sample could be generated by taking a stratified random sample by taking a simple random sample from each of the 20 campuses and again asking each student in the sample to report the distance from their hometown to campus. Certain problems might arise with self-reporting of distances, such as recording error or poor recall. This study is enumerative because there exists a finite, identifiable population of objects from which to sample.

7 **a** Number of observations equalled $2 \times 2 \times 2 = 8$

b This could be called an analytic study because the data would be collected on an existing process. There is no sampling frame.

Chapter 1

Section 1.2

9 **a**

Stem	Leaves
2	2 3
3	2 3 4 4 5 6 7 7 8 9
4	0 1 3 5 6 8 8 9
5	0 0 0 0 1 1 1 4 4 5 5 6 6 6 7 8 9
6	0 0 0 0 1 2 2 2 2 3 3 4 4 4 5 6 6 6 7 7 8 9 9 9 9
7	0 0 0 1 2 2 3 3 4 5 5 5 5 5 6 6 8
8	0 2 2 3 3 4 4 8
9	0 1 2 2 3 3 3 3 5 6 6 6 7 8 8
10	2 3 4 4 4 5 5 6 8 8
11	2 3 3 5 9 9 9
12	3 7
13	8
14	3 6
15	0 0 3 5
16	
17	
18	9

stem: ones
leaf: tenths

b Possible answers would be the most frequently occurring values (those appearing with a stem of 6) or the middle value after the data is ranked (7.0).

c Fairly spread out

d The distribution is not symmetric. It is positively skewed.

e There appears to be one outlier, 18.9.

11

Class	Frequency	Relative Frequency
4000 –< 4200	1	.01
4200 –< 4400	2	.02
4400 –< 4600	9	.09
4600 –< 4800	13	.13
4800 –< 5000	18	.18
5000 –< 5200	22	.22
5200 –< 5400	20	.20
5400 –< 5600	7	.07
5600 –< 5800	7	.07
5800 –< 6000	1	.01
	100	1.00

Chapter 1

13 **a**

Class	Frequency	Relative Frequency	Density
.15 -< .25	8	.02192	.2192
.25 -< .35	14	.03836	.3836
.35 -< .45	28	.07671	.7671
.45 -< .50	24	.06575	1.3150
.50 -< .55	39	.10685	2.1370
.55 -< .60	51	.13973	2.7945
.60 -< .65	106	.29041	5.8082
.65 -< .70	84	.23014	4.6027
.70 -< .75	11	.03014	.6027
	n = 365	1.00001	

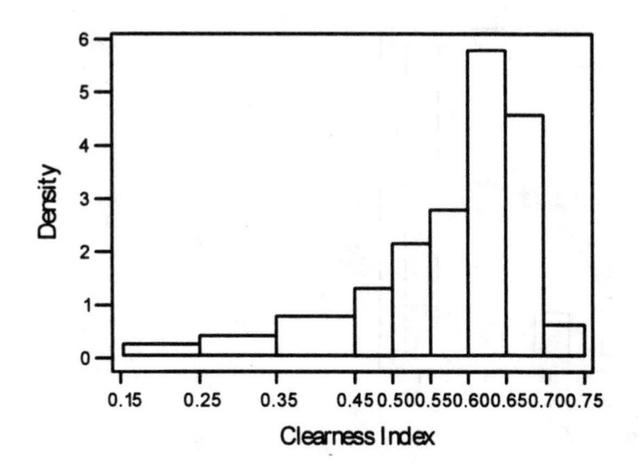

b $\frac{8+14}{365} = .06$ or 6%

or $.02192 + .03836 = .06$ or 6%

c $\frac{84+11}{365} = .26$ or 26%

or $.23014 + .03014 = .26028$ or 26%

15

Category	Frequency	Relative Frequency
J	10	.1667
F	9	.1500
B	7	.1167
M	4	.0667
C	3	.0500
N	6	.1000
O	21	.3500
	n = 60	1.0001

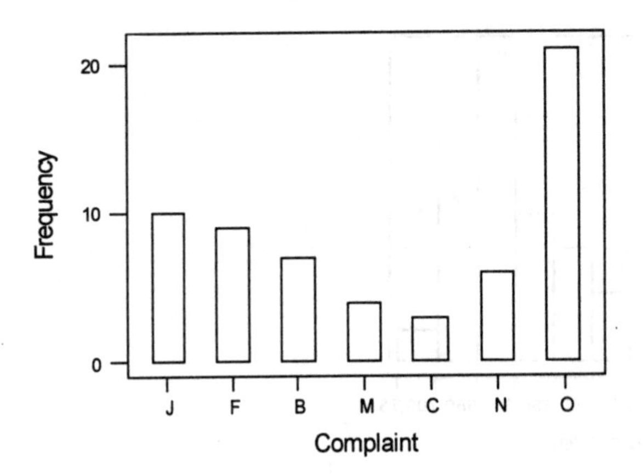

Chapter 1

17 **a**

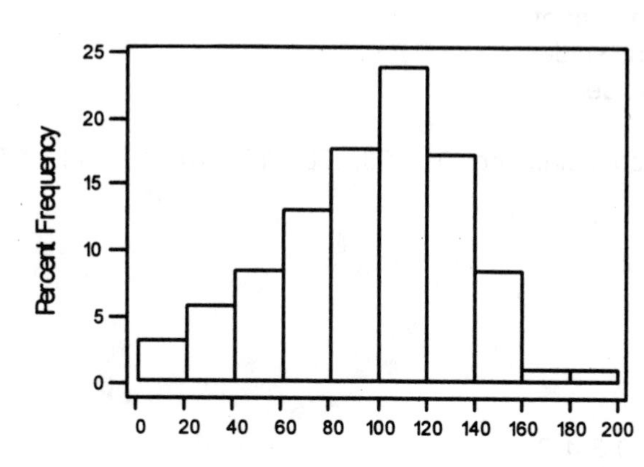

b Estimate = sample proportion = $\frac{(46+33+16+2+2)}{191} = \frac{99}{191} = .518$

c The relative frequency for 40 –< 60 is .084; suppose that half of this (.042) is between 50 and 60. Similarly, suppose that 25% of the .173 relative frequency for 120 –< 140 (.043) is between 120 and 125. The estimate is then $.042+.131+.178+.241+.043 = .635$.

19

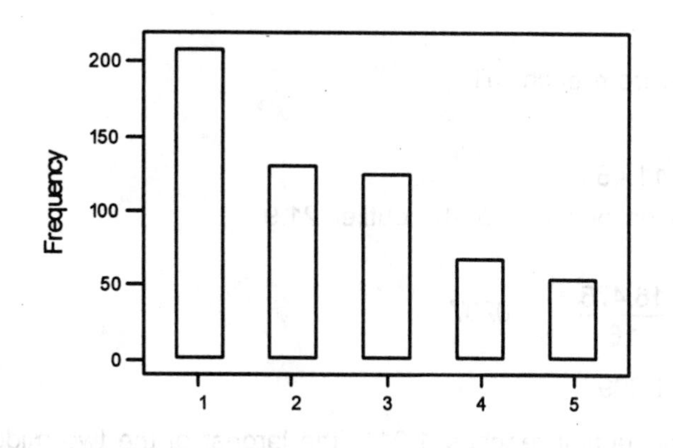

Chapter 1

1. incorrect component
2. missing component
3. failed component
4. insufficient solder
5. excess solder

21 The rounded relative frequencies are .01, .03, .08, .17, .19, .19, .11, .17, .03, .02, and .01, which sum to 1.01.

23

6L	4 3 0
6H	7 6 9 6 8 9
7L	4 2 0 1 4 2 0 2
7H	
8L	0 1 1 2 1 1 4 1 0 3 4 2
8H	9 5 9 5 5 7 8
9L	3 0
9H	5 8

This display brings out the gap in the data — there are no scores in the high 70's.

Section 1.3

25 **a** $\bar{x} = 192.57$, $\tilde{x} = 189.0$; the mean is larger than the median, but they are still fairly close together

b $\bar{x} = 189.71$, $\tilde{x} = 189.0$; the mean is lowered, the median stays the same

27 $\bar{x}_{tr} = 191.0$

$\dfrac{1}{14} = .0714$ or 7.14% trimmed from each tail

29 $\bar{x} = 12.01$, $\tilde{x} = 11.35$, $\bar{x}_{tr(10)} = 11.46$

$\bar{x}_{tr(10)}$ or $\tilde{x}$ would be good choices because of the outlier 21.9

31 **a** $\Sigma x_i = 16.475$, so $\bar{x} = \dfrac{16.475}{16} = 1.0297$

$\tilde{x} = \dfrac{(1.007 + 1.011)}{2} = 1.009$

b 1.394 can be decreased until it reaches 1.011 (the largest of the two middle values) — i.e. by $1.394 - 1.011 = .383$. If it is decreased by more than .383, the median will change.

33 **a** $\frac{7}{10} = .7$

b $\bar{x} = .7$ = proportion of successes

c $\frac{s}{25} = .80$ thus $s = (0.80)(25) = 20$

total of 20 successes

$20 - 7 = 13$ of the new cars would have to be successes

35 median $= \frac{(57+79)}{2} = 68.0$, 20% trimmed mean = 66.2,

30% trimmed mean = 67.5

Section 1.4

37 **a** $\bar{x} = \frac{320}{8} = 40.0$ so the deviations are 0, 3, −1, −5, −3, 3, 6, −3 and

$s^2 = \frac{98}{7} = 14.00$, $s = 3.74$

b $\sum x_i = 320$, $\sum x_i^2 = 12{,}898$, $s^2 = \frac{[12898 - (320)^2/8]}{7} = \frac{98}{7} = 14.00$,

$s = 3.74$

c New values are 5, 8, 4, 0, 2, 8, 11, 2, sum = 40, sum of squares = 298, so new

$$s^2 = \frac{[298 - (40)^2/8]}{7} = \frac{298 - 200}{7} = \frac{98}{7} = 14.00 \text{ as before.}$$

d With y_i = new value, each $y_i = 10x_i$, so s^2 for

y_i's $= 100 \times (s^2$ for x_i's$) = 1400$, s for y_i's $= 10 \times (s$ for x_i's$) = 37.4$

39 **a** $\sum x = 2.75 + \ldots + 3.01 = 56.80$, $\sum x^2 = (2.75)^2 + \ldots + (3.01)^2 = 197.8040$

b $s^2 = \frac{[197.8040 - (56.80)^2/17]}{16} = \frac{8.0252}{16} = .5016$, $s = .708$. Roughly speaking, the magnitude of a typical deviation from the sample mean is about .7.

41 New values are 4.0, 4.8, 1.6, 8.0, 3.8, .8, 4.2, .2, 2.4, 1.4

sum = 31.2, sum of squares = 146.08, $s^2 = \frac{[146.08 - (31.2)^2/10]}{9} = 5.4151$, $s = 2.327$; subtracting any other value would result in the same value of s.

43 a $\sum x_i^2 - \frac{(\sum x_i)^2}{n} = 17.371567 - \frac{(16.475)^2}{16} = .407465$, so

$s^2 = \frac{.407465}{15} = .027164$ and $s = .164816$

b lower fourth $= \frac{(.913 + .915)}{2} = .914$, upper fourth $= \frac{(1.132 + 1.140)}{2} = 1.136$, and $f_s = 1.136 - .914 = .222$

c lower fourth $-1.5f_s = .581$ and upper fourth $+1.5f_s = 1.469$. Since there are no observations either below .581 or above 1.469, the data set contains no outliers. The left edge of the box is above .914 and the lower whisker extends from that edge out to .736, the smallest sample x. Similarly, the box's right edge is above 1.136 and the right whisker extends out to 1.394.

45 Let d denote the fifth deviation. Then $.3 + .9 + 1.0 + 1.3 + d = 0$, i.e. $3.5 + d = 0$, so $d = -3.5$. One sample for which these are the deviations is $x_1 = 3.8$, $x_2 = 4.4$, $x_3 = 4.5$, $x_4 = 4.8$, and $x_5 = 0$ (obtained by adding 3.5 to each deviation; adding any other number will produce a different sample with the desired property).

47 The lower fourth, median, and upper fourth are 104, 132, and 152.5, respectively, so $f_s = 48.5$ and $1.5f_s = 72.75$. Since all observations are between $104 - 72.75$ and $152.5 + 72.75$, there are no outliers. The lower and upper whiskers extend to 87 and 211, respectively, the smallest and largest sample x_i's. The median is a bit closer to the upper fourth than to the lower fourth, suggesting a very mild left skew in the middle half of the data. However, the upper whisker is much longer than the lower whisker, indicating a much longer upper than lower tail — a substantial positive skew.

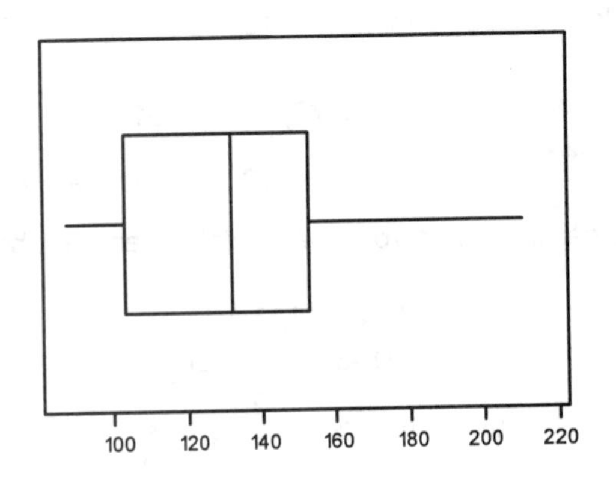

49 Divide the new s^2 by $(10)^2$.

51

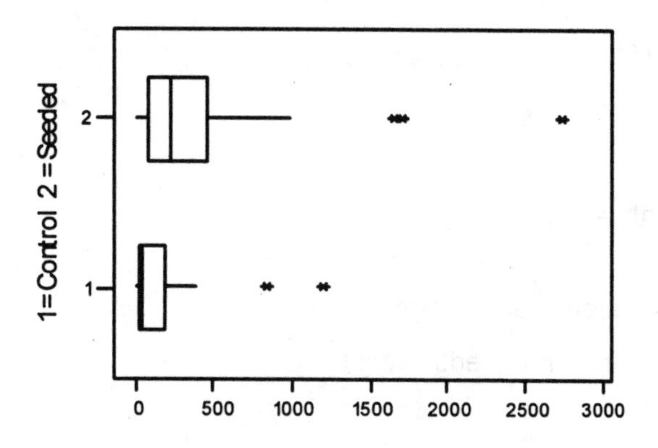

Both boxplots show evidence of substantial positive skewness, and there are several outliers in each sample. The plot for seeded data is much more stretched out than is the plot for control data.

Supplementary

53 **a** mode = 62
b mode = category with highest frequency

55 $\sum x_i = 163.2$

$$100\left(\frac{1}{15}\right)\%\text{ trimmed mean} = \frac{163.2 - 8.5 - 15.6}{13} = 10.70$$

$$100\left(\frac{2}{15}\right)\%\text{ trimmed mean} = \frac{163.2 - 8.5 - 8.8 - 15.6 - 13.7}{11} = 10.60$$

$$\therefore \frac{1}{2}\bullet(100)\left(\frac{1}{15}\right) + \frac{1}{2}\bullet(100)\left(\frac{2}{15}\right) = 100\left(\frac{1}{10}\right) = 10\%\text{ trimmed mean is}$$

$$\frac{1}{2}(10.70) + \frac{1}{2}(10.60) = 10.65$$

Chapter 1

57 **a** $\bar{y} = \frac{\sum y_i}{n} = \frac{\sum (ax_i + b)}{n} = \frac{a\sum x_i + b}{n} = a\bar{x} + b$

$$s_y^2 = \frac{\sum (y_i - \bar{y})^2}{n-1} = \frac{\sum (ax_i + b - (a\bar{x} + b))^2}{n-1} = \frac{\sum (ax_i - a\bar{x})^2}{n-1}$$

$$= \frac{a^2 \sum (x_i - \bar{x})^2}{n-1} = a^2 s_x^2$$

b 189.14, 1.87

59 $\bar{x} = .9255$, $s = .0809$, $\tilde{x} = .93$
lower fourth = .855, upper fourth = .96

0.7	8	stem: tenths digit
0.8	1 1 5 5 6	leaf: hundredths digit
0.9	2 2 3 3 3 3 5 5 6 6	
1.0	0 5 6 6	

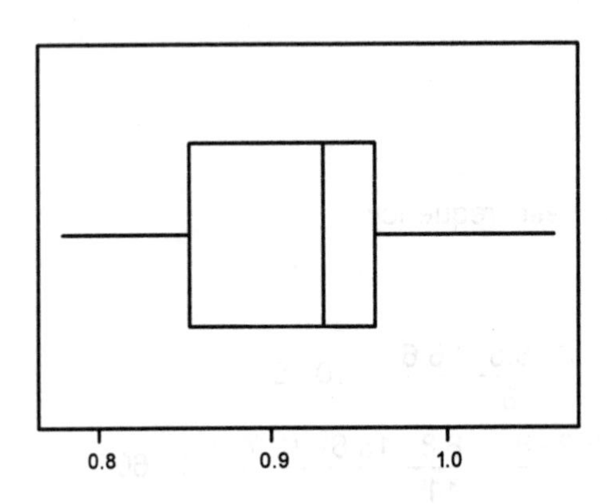

The data appears to be a bit skewed toward smaller values (negatively skewed). There are no outliers. The mean and the median are close in value.

61

	$\bar{x}$	$\tilde{x}$	lower fourth	upper fourth	s
Smoking	693.364	693	609.5	760.0	103.864
Nonsmoking	961.091	947	916.5	977.0	113.810

Chapter 1

Smoking		Nonsmoking
98 93 45	5	
93 55 21	6	
93 67 53 14	7	45
95	8	99
	9	03 30 45 47 61 73 81
	10	86
	11	
	12	02

stem: hundreds
leaf: tens and ones

smoking: $f_s = 150.5$

$$609.5 - (1.5)(150.5) = 383.75$$
$$760 + (1.5)(150.5) = 985.75$$

no outliers

nonsmoking: $f_s = 60.5$

$$916.5 - (1.5)(60.5) = 825.75$$
$$(977) + (1.5)(60.5) = 1{,}067.75$$
$$916.5 - (3)(60.5) = 735$$
$$977 + (3)(60.5) = 1{,}158.5$$

1=Smoking 2=Nonsmoking

2
1

500 600 700 800 900 1000 1100 1200

The milk volume in gallons per day appears to be greater in nonsmoking than in smoking mothers. The smallest value for nonsmoking mothers exceeds the median value for smoking mothers. There is much less spread in the middle 50% for nonsmoking mothers, which results in the identification of more outliers in the nonsmoking group.

63 **a** Since the constant $\overline{x}$ is subtracted from each x_i to obtain y_i, and addition or subtraction of a constant doesn't affect variability, $s_y^2 = s_x^2$ and $s_y = s_x$.

b Let $c = \frac{1}{s}$ where s is the sample s.d. of the x_i's and also (by a) of the y_i's.

Then $s_z = cs_y = \left(\frac{1}{s}\right)s = 1$, and $s_z^2 = 1$. That is, the "standardized" quantities $z_1, \ldots, z_n$ have a sample variance and s.d. of 1.

65

Stem	Leaves
1H	96, 90, 88, 86
2L	43, 36
2H	87, 99, 92, 91, 92
3L	24, 00, 46, 05, 04, 34, 15, 35, 32, 49, 24, 01, 33, 03, 18, 06, 45, 45, 37, 20, 29
3H	99, 93, 75, 99, 67, 53, 57, 54, 61
4L	11, 48, 19, 47, 19, 08, 41, 28, 12
4H	51, 98, 89, 68
5L	
5H	85

stem = tenths
leaves = hundredths and thousandths

Class	Frequency	Relative Frequency
.150 –< .200	4	.073
.200 –< .250	2	.036
.250 –< .300	5	.091
.300 –< .350	21	.382
.350 –< .400	9	.164
.400 –< .450	9	.164
.450 –< .500	4	.073
.500 –< .550	0	.000
.550 –< .600	1	.018
	55	1.001

$\bar{x} = .3489$, $s = .0800$, $\tilde{x} = .3370$, $f_s = .4035 - .3035 = .1000$, $.4035 + 1.5f_s = .5535$, so .585 is a mild outlier. The distribution of values is reasonably symmetric, with a high peak in the low .3's.

67 **a** When there is perfect symmetry, the smallest observation y_1 and the largest observation y_n will be equidistant from $\tilde{x}$, so $y_n - \bar{x} = \bar{x} - y_1$. Similarly, the second smallest and second largest will be equidistant from $\tilde{x}$, so $y_{n-1} - \bar{x} = \bar{x} - y_2$, and so on. Thus the first and second numbers in each pair will be equal, so that each point in the plot will fall exactly on the 45 degree line. When the data is positively skewed, y_n will be much further from $\tilde{x}$ than is y_1, so $y_n - \tilde{x}$ will considerably exceed $\tilde{x} - y_1$ and the point $(y_n - \tilde{x}, \tilde{x} - y_1)$ will fall considerably below the 45 degree line. A similar comment applies to other points in the plot.

b The first point in the plot is $(y_n - \tilde{x}, \tilde{x} - y_1) = (2745.6 - 221.6, 221.6 - 4.1)$ = (2524.0, 217.5). The others are (1476.2, 213.9), (1434.4, 204.1), (756.4, 190.2), (481.8, 188.9), (267.5, 181.0), (208.4, 129.2), (112.5, 106.3), (81.2, 103.3), (53.1, 102.6), (53.1, 92.0), (33.4, 23.0) and (20.9, 20.9). The first number in each of the first seven pairs greatly exceeds the second number, so each point falls well below the 45 degree line. A substantial positive skew (stretched upper tail) is indicated.

CHAPTER 2

Section 2.1

1 **a** {(A,A),(A,E),(A,J),(E,A),(E,E),(E,J),(J,A),(J,E),(J,J)}, where the first component of a pair identifies the country in which the older car was manufactured and the second component identifies the country in which the newer car was manufactured.

b {(A,E),(A,J),(E,A),(J,A)}

c {(A,E),(A,J),(E,A),(E,E),(E,J),(J,A),(J,E),(J,J)}
Complement = {(A,A)}, which is a simple event.

3

Outcome Number	Outcome
1	1 1 1
2	1 1 2
3	1 1 3
4	1 2 1
5	1 2 2
6	1 2 3
7	1 3 1
8	1 3 2
9	1 3 3
10	2 1 1
11	2 1 2
12	2 1 3
13	2 2 1
14	2 2 2
15	2 2 3
16	2 3 1
17	2 3 2
18	2 3 3
19	3 1 1
20	3 1 2
21	3 1 3
22	3 2 1
23	3 2 2
24	3 2 3
25	3 3 1
26	3 3 2
27	3 3 3

Chapter 2

b Outcome numbers 1, 14, 27
c Outcome numbers 6, 8, 12, 16, 20, 22
d Outcome numbers 1, 3, 7, 9, 19, 21, 25, 27

5 **a** {11, 22, 33}
b {1213, 1312, 1231, 1321, 2123, 2132, 2312, 2321, 3123, 3132, 3213, 3231}

7 **a** S = { BBBAAAA, BBABAAA, BBAABAA, BBAAABA, BBAAAAB,
BABBAAA, BABABAA, BABAABA, BABAAAB, BAABBAA,
BAABABA, BAABAAB, BAAABBA, BAAABAB, BAAAABB,
ABBBAAA, ABBABAA, ABBAABA, ABBAAAB, ABABBAA,
ABABABA, ABABAAB, ABAABBA, ABAABAB, ABAAABB,
AABBBAA, AABBABA, AABBAAB, AABABBA, AABABAB,
AABAABB, AAABBBA, AAABBAB, AAABABB, AAAABBB}
b { AAAABBB, AAABABB, AAABBAB, AABAABB, AABABAB}

9 **a**

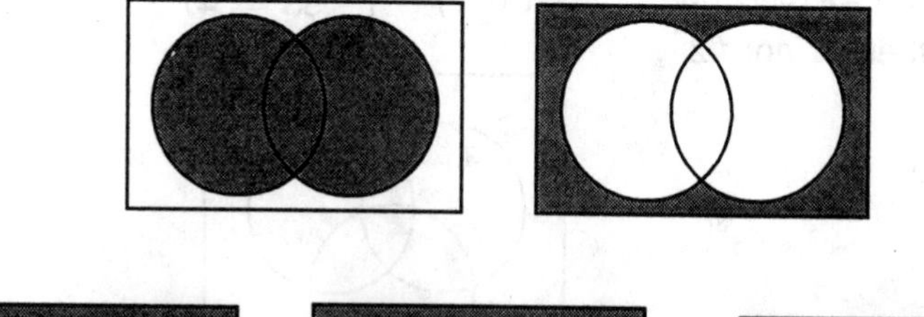

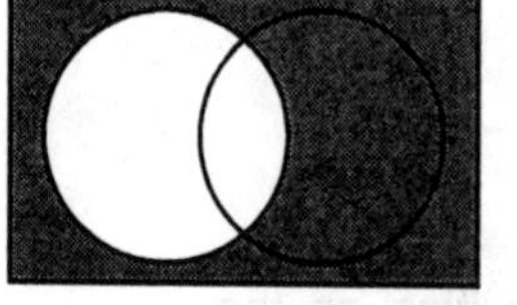

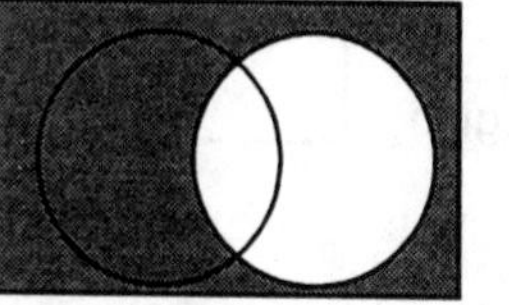

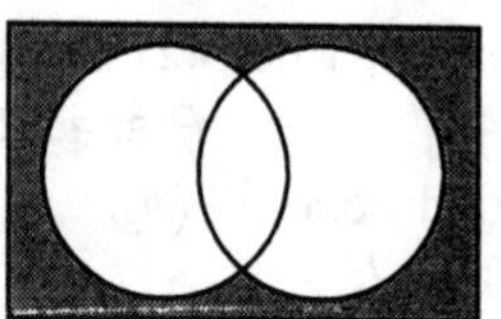

b

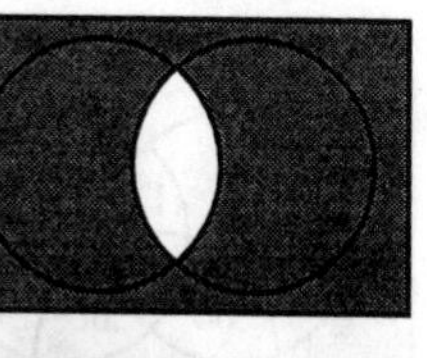

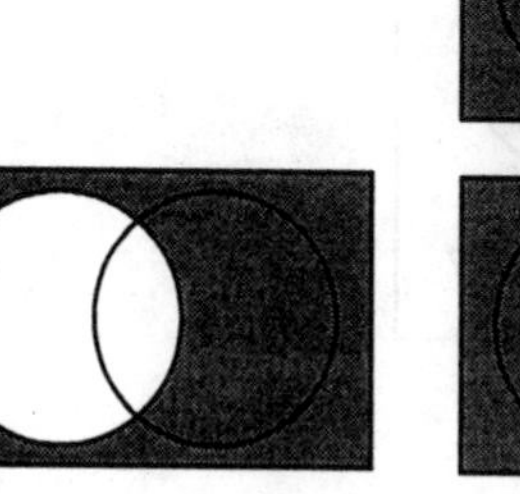

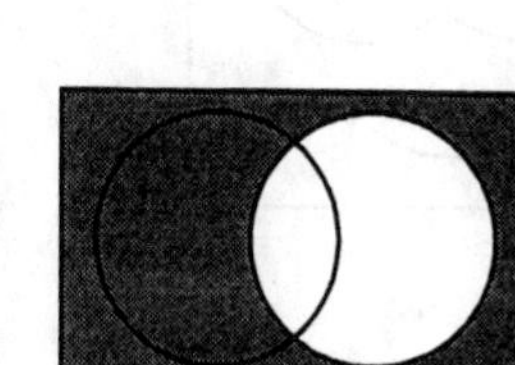

Chapter 2

Section 2.2

11 **a** .07
b .30
c .57

13 **a** awarded either #1 or #2 (or both);

$P(A_1 \cup A_2) = P(A_1) + P(A_2) - P(A_1 \cap A_2) = .22 + .25 - .11 = .36$

b awarded neither #1 nor #2;

$P(A_1' \cap A_2') = P((A_1 \cup A_2)') = 1 - P(A_1 \cup A_2) = 1 - .36 = .64$

c awarded at least one of #1, #2, #3; $P(A_1 \cup A_2 \cup A_3)$

$= P(A_1) + P(A_2) + P(A_3) - P(A_1 \cap A_2) - P(A_1 \cap A_3)$
$\quad - P(A_2 \cap A_3) + P(A_1 \cap A_2 \cap A_3)$

$= .22 + .25 + .28 - .11 - .05 - .07 + .01 = .53$

d awarded none of the three projects;

$P(A_1' \cap A_2' \cap A_3') = 1 - P(\text{awarded at least one}) = 1 - .53 = .47$

e awarded #3 but neither #1 nor #2;

$P(A_1' \cap A_2' \cap A_3)$
$= P(A_3) - P(A_1 \cap A_3)$
$\quad - P(A_2 \cap A_3) + P(A_1 \cap A_2 \cap A_3)$
$= .28 - .05 - .07 + .01$
$= .17$

f either (neither #1 nor #2) or #3:

$P((A_1' \cap A_2') \cup A_3) = P(\text{shaded region})$
$= P(\text{awarded none}) + P(A_3)$
$= .47 + .28 = .75$

Alternatively, answers to **a – f** can be obtained from probabilities on the accompanying Venn diagram:

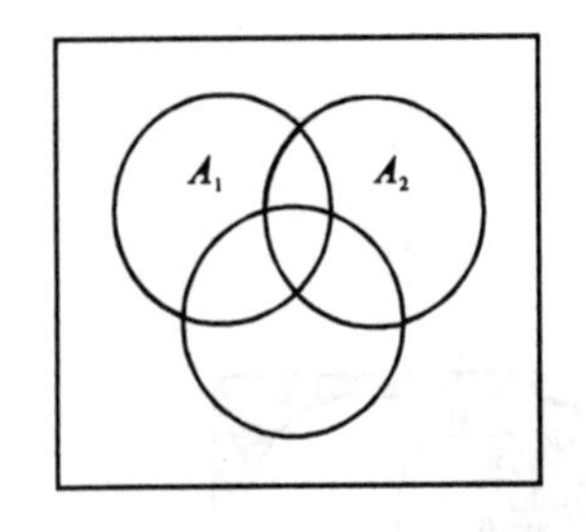

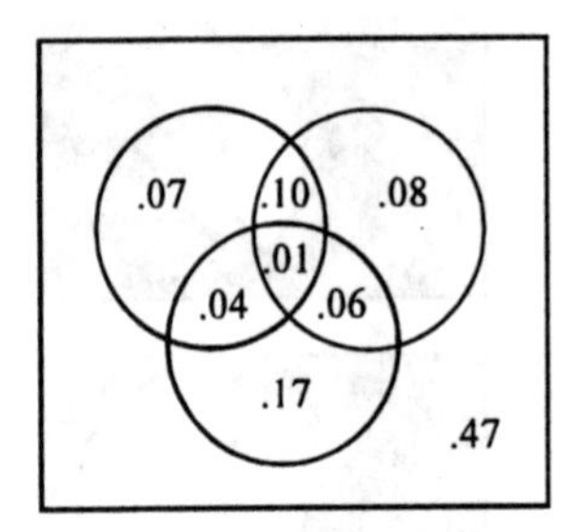

Chapter 2

15 **a** Let event E be the event that at most one purchases an electric dryer. Then E' is the event that at least two purchase electric dryers.
Thus, $P(E') = 1 - P(E) = 1 - .087 = .913$.

b Let event A be the event that all five purchase gas. Let event B be the event that all five purchase electric. All other possible outcomes are those in which at least one of each type is purchased. Thus, the desired probability =
$1 - P(A) - P(B) = 1 - .0768 - .0102 = .913$.

17 **a** $P(A') = 1 - P(A) = 1 - .6 = .4$

b $P(A \cup B) = P(A) + P(B) = .6 + .15 = .75$

c $P(A' \cap B') = P((A \cup B)') = 1 - P(A \cup B) = 1 - .75 = .25$

19 **a** {(M,2)},{(M,4)},{(Q,2)},{(Q,4)}

b .32

c $P(M) = .25 + .16 = .41$

21 **a** .10

b $P(\text{low auto}) = P(\{(L,N),(L,L),(L,M),(L,H)\}) = .04 + .06 + .05 + .03 = .18$
$P(\text{low homeowner's}) = .06 + .10 + .03 = .19$

c $P(\text{same deductible for both}) = P(\{(L,L),(M,M),(H,H)\}) = .06 + .20 + .15 = .41$

d $P(\text{deductibles are different}) = 1 - P(\text{same deductibles}) = 1 - .41 = .59$

e $P(\text{at least one low deductible}) = P(\{(L,N),(L,L),(L,M),(L,H),(M,L),(H,L)\})$
$= .04 + .06 + .05 + .03 + .10 + .03 = .31$

f $P(\text{neither deductible low}) = 1 - P(\text{at least one low}) = 1 - .31 = .69$

23 $P(\text{at least two must be selected}) = 1 - P(\text{first one is blank})$
complementary events

$$= 1 - \frac{15}{25} = \frac{10}{25} = \frac{2}{5} = .40$$

25 $B = A \cup (B \cap A')$, so
$P(B) = P(A) + P(B \cap A')$ (since A and $(B \cap A')$
$\geq P(A)$ are mutually exclusive)
since $P(B \cap A')$ is ≥ 0.

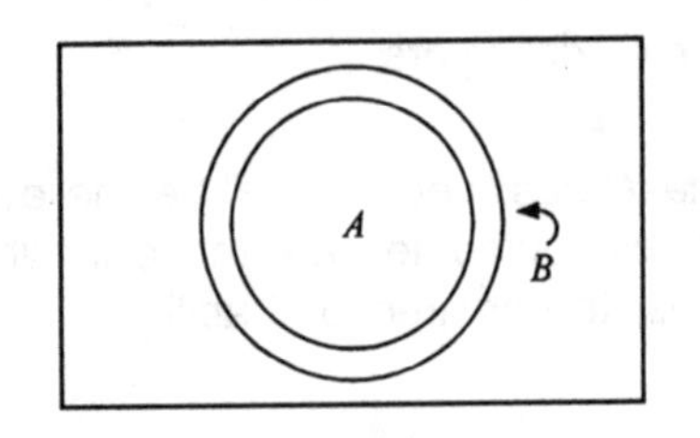

27 $P(A \cap B) = P(A) + P(B) - P(A \cup B) + .65$

$P(A \cap C) = .55,\ P(B \cap C) = .60$

$$P(A \cap B \cap C) = P(A \cup B \cup C) - P(A) - P(B) - P(C) + P(A \cap B) + P(A \cap C) + P(B \cap C)$$

$$= .95 - .7 - .8 - .75 + .65 + .55 + .60$$

$$= .53$$

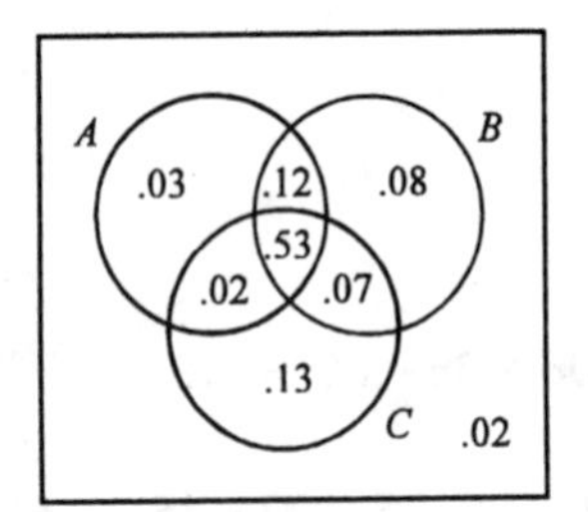

a $P(A \cup B \cup C) = .98$ as given

b $P(\text{none selected}) = 1 - P(A \cup B \cup C) = .02$

c $P(\text{only a radio is selected}) = .13$ from the Venn diagram

d $P(\text{exactly one of the three}) = .03 + .08 + .13 = .24$

29 There are 27 equally likely outcomes.

a $P(\text{all the same}) = P((1,1,1) \text{ or } (2,2,2) \text{ or } (3,3,3)) = \dfrac{3}{27} = \dfrac{1}{9}$

b $P(\text{at most two are assigned to the same station}) = 1 - P(\text{all three are the same})$

$$= 1 - \frac{3}{27} = \frac{24}{27} = \frac{8}{9}$$

c $P(\text{all different}) = P(\{(1,2,3),(1,3,2),(2,1,3),(2,3,1),(3,1,2),(3,2,1)\})$

$$= \frac{6}{27} = \frac{2}{9}$$

Chapter 2

Section 2.3

31 **a** $10 \times 9 \times 8 = 720$

b $\binom{10}{3} = \frac{10!}{3!7!} = 120$

c P(all three are new) = # of ways of visiting all new/# of ways of visiting

$$= \frac{\binom{4}{3}}{\binom{10}{3}} = \frac{4}{120} = \frac{1}{30}$$

33 **a** $7 \times 7 \times 7 \times 7 = 2401$

b $P(\text{all are news programs}) = \frac{\text{number of all-news sequences}}{2401}$

$$= \frac{3 \times 5 \times 4 \times 2}{2401} = \frac{120}{2401} = .050$$

c $P(\text{at least two are news}) = 1 - P(\text{at most one is news})$

$$= 1 - P(\text{none are news}) - P(\text{exactly one is news}),$$

$P(\text{none are news}) = \frac{4 \times 2 \times 3 \times 5}{2401} = \frac{120}{2401} = .050$

$P(\text{exactly one is news}) = P(\text{only first is news}) + P(\text{only second is news})$
$+ P(\text{only third is news}) + P(\text{only fourth is news})$

$$= \frac{3 \times 2 \times 3 \times 5}{2401} + \frac{4 \times 5 \times 3 \times 5}{2401} + \frac{4 \times 2 \times 4 \times 5}{2401} + \frac{4 \times 2 \times 3 \times 2}{2401}$$

$$= \frac{90 + 300 + 160 + 48}{2401} = \frac{598}{2401} = .249.$$

$\therefore P(\text{at least two are news}) = 1 - .050 - .249 = .701$

d P(6:00 news and 6:30 news)

= P(5:30 any show, 6:00 news, 6:30 news, 7:00 any show)

$$= \frac{7 \times 5 \times 4 \times 7}{2401}$$

$$= \frac{980}{2401}$$

$$= .408$$

35 **a** $\binom{20}{5} = \frac{20!}{5!15!} = 15{,}504$

b $\binom{8}{4} \cdot \binom{12}{1} = 840$

Chapter 2

c $P(\text{exactly four have cracks}) = \dfrac{\binom{8}{4}\binom{12}{1}}{\binom{20}{5}} = \dfrac{840}{15{,}504} = .0542$

d $P(\text{at least four}) = P(\text{exactly four}) + P(\text{exactly five})$

$$= .0542 + \frac{\binom{8}{5}\binom{5}{0}}{\binom{20}{5}} = .0578$$

37 There are ten possible outcomes — $\binom{5}{2}$ ways to select the positions for B's votes: BBAAA, BABAA, BAABA, BAAAB, ABBAA, ABABA, ABAAB, AABBA, AABAB, and AAABB. Only the last two have A ahead of B throughout the vote count. Since the outcomes are equally likely, the desired probability is $\dfrac{2}{10} = .20$.

39 There are $\binom{60}{5}$ ways to select the 5 runs. Each catalyst is used in 12 different runs, so the number of ways of selecting one run from each of these 5 groups is 12^5. Thus the desired probability is $\dfrac{12^5}{\binom{60}{5}} = .0456$.

41 **a** If the A's are distinguishable from one another, and similarly for the B's, C's, and D's, then there are 12! possible chain molecules. Six of these are

$$A_1A_2A_3B_2C_3C_1D_3C_2D_1D_2B_3B_1,\ A_1A_3A_2B_2C_3C_1D_3C_2D_1D_2B_3B_1$$
$$A_2A_1A_3B_2C_3C_1D_3C_2D_1D_2B_3B_1,\ A_2A_3A_1B_2C_3C_1D_3C_2D_1D_2B_3B_1$$
$$A_3A_1A_2B_2C_3C_1D_3C_2D_1D_2B_3B_1,\ A_3A_2A_1B_2C_3C_1D_3C_2D_1D_2B_3B_1$$

These 6 (= 3!) differ only with respect to ordering of the 3 A's. In general, groups of 6 chain molecules can be created such that within each group only the ordering of the A's is different. When the A subscripts are suppressed, each group of 6 "collapses" into a single molecule (B's, C's, and D's are still distinguishable). At this point there are thus $\dfrac{12!}{3!}$ molecules. Now suppressing subscripts on the B's, C's, and D's in turn gives ultimately $\dfrac{12!}{(3!)^4} = 369{,}600$ chain molecules.

b Think of the group of 3 A's as a single entity, and similarly for the B's, C's, and D's. Then there are 4! ways to order these entities, and thus 4!

molecules in which the A's are contiguous, the B's are also, the C's are also, and the D's are also. Thus $P(\text{all together}) = \dfrac{4!}{369{,}600} = .00006504$.

43

seats: 1 2 3 4 5 6

$P(J \,\&\, P \text{ in } 1 \,\&\, 2) = \dfrac{2\times1\times4\times3\times2\times1}{6\times5\times4\times3\times2\times1} = \dfrac{1}{15} = .067$

$P(J \,\&\, P \text{ next to one another}) = P(J \,\&\, P \text{ in } 1 \,\&\, 2) + \ldots + P(J \,\&\, P \text{ in } 5 \,\&\, 6)$

$= 5\times\dfrac{1}{15} = \dfrac{1}{3} = .333$

$P(\text{at least one } H \text{ next to his } W) = 1 - P(\text{no } H \text{ next to his } W)$

We count the # of ways of no H next to his W as follows:

# of orderings with a H–W pair in seats #1 & 3 and no H next to his $W = 6\times4\times1\times2\times1\times1 = 48$

↓ ↙ ↓

pair   can't put the mate of seat #2 here, else H-W in #5, 6

# of orderings without a H–W pair in seats #1 & 3 and no H next to his $W = 6\times4\times2\times2\times2\times1 = 192$

↓

can't be mate of person in seat #1 or #2

$\therefore$ # of seating arrangements with no H next to $W = 48 + 192 = 240$

$\therefore P(\text{no } H \text{ next to } W) = \dfrac{240}{6\times5\times4\times3\times2\times1} = \dfrac{1}{3}$, so

$P(\text{at least one } H \text{ next to his } W) = 1 - \dfrac{1}{3} = \dfrac{2}{3} = .667$

45 $\dbinom{n}{k} = \dfrac{n!}{k!(n-k)!} = \dfrac{n!}{(n-k)!k!} = \dbinom{n}{n-k}$

# of subsets of size k = # of subsets of size $n-k$, because to each subset of size k there corresponds exactly one subset of size $n-k$ (the $n-k$ objects not in the subset of size k).

Section 2.4

47 A - individual is tall

B - individual plays professional basketball

note - $P(B) << P(A)$

$P(A \cap B)$ is very close to $P(B)$ (both are small)

$$P(A|B) = \frac{P(A \cap B)}{P(B)} = \frac{\text{very small number}}{\text{very small number}} = \text{high probability}$$

$$P(B|A) = \frac{P(A \cap B)}{P(A)} = \frac{\text{very small number}}{\text{large number}} = \text{low probability}$$

The probability that someone selected at random is over 6 feet tall given that they play professional basketball is much greater than the probability that someone plays professional basketball given that they are over 6 feet tall.

49 $$P(GG|\text{at least one } G) = \frac{P(GG \cap \text{at least one G})}{P(\text{at least one } G)}$$

$$= \frac{P(GG)}{P(\text{at least one } G)} = \frac{\left[\frac{4}{38 \times 38}\right]}{1 - P(\text{neither } G)}$$

$$= \frac{\left[\frac{4}{1444}\right]}{\left[1 - \frac{36 \times 36}{38 \times 38}\right]} = .027$$

51

a $P(M \cap LS \cap Pr) = .05$ directly from the table of probabilities

b $P(M \cap Pr) = P((M,Pr,LS)) + P((M,Pr,SS)) = .05 + .07 = .12$

c $P(SS)$ = sum of 9 probabilities in SS table = .56, $P(LS) = 1 - .56 = .45$

d $P(M) = .08 + .07 + .12 + .10 + .05 + .07 = .49$,
$P(Pr) = .02 + .07 + .07 + .02 + .05 + .02 = .25$

e $$P(M|SS \cap Pl) = \frac{P(M \cap SS \cap Pl)}{P(SS \cap Pl)} = \frac{.08}{.04 + .08 + .03} = .533$$

f $$P(SS|M \cap Pl) = \frac{P(SS \cap M \cap Pl)}{P(M \cap Pl)} = \frac{.08}{.08 + .10} = .444,$$
$P(LS|M \cap Pl) = 1 - P(SS|M \cap Pl) = 1 - .444 = .556$

53 Let A_1 be the event that #1 fails and A_2 be the event that #2 fails. We assume that $P(A_1) = P(A_2) = q$ and that $P(A_1|A_2) = P(A_2|A_1) = r$. Then one approach is as follows:

$P(A_1 \cap A_2) = P(A_2|A_1) \bullet P(A_1) = rq = .01$

$P(A_1 \cup A_2) = P(A_1 \cap A_2) + P(A_1' \cap A_2) + P(A_1 \cap A_2') = rq + 2(1 - r)q = .07$

These two equations give $2q - .01 = .07$, from which $q = .04$ and $r = .25$.
Alternatively, with $t = P(A_1' \cap A_2) = P(A_1 \cap A_2')$, $t + .01 + t = .07$, implying $t = .03$ and thus $q = .04$ without reference to conditional probability.

Chapter 2

55 $P(A_1) = .22,\ P(A_2) = .25,\ P(A_3) = .28,\ P(A_1 \cap A_2) = .11,\ P(A_1 \cap A_3) = .05,$
$P(A_2 \cap A_3) = .07,\ P(A_1 \cap A_2 \cap A_3) = .01$

a $$P(A_2|A_1) = \frac{.11}{.22} = .5$$

b $$P(A_2 \cap A_3|A_1) = \frac{P(A_1 \cap A_2 \cap A_3)}{P(A_1)} = \frac{.01}{.22} = .0455$$

c $$P(A_2 \cup A_3|A_1) = \frac{P((A_2 \cup A_3) \cap A_1)}{P(A_1)} = \frac{P((A_1 \cap A_2) \cup (A_1 \cap A_3))}{.22}$$
$$= \frac{P(A_1 \cap A_2) + P(A_1 \cap A_3) - P(A_1 \cap A_2 \cap A_3)}{.22} = \frac{.15}{.22} = .682$$

d $$P(A_1 \cap A_2 \cap A_3|A_1 \cup A_2 \cup A_3) = \frac{P(A_1 \cap A_2 \cap A_3)}{P(A_1 \cup A_2 \cup A_3)} = \frac{.01}{.53} = .0189.$$

This is the probability of being awarded all three projects given that at least one project was awarded.

57 $$P(A|B) + P(A'|B) = \frac{P(A \cap B)}{P(B)} + \frac{P(A' \cap B)}{P(B)} = \frac{P(A \cap B) + P(A' \cap B)}{P(B)} = \frac{P(B)}{P(B)} = 1$$

59

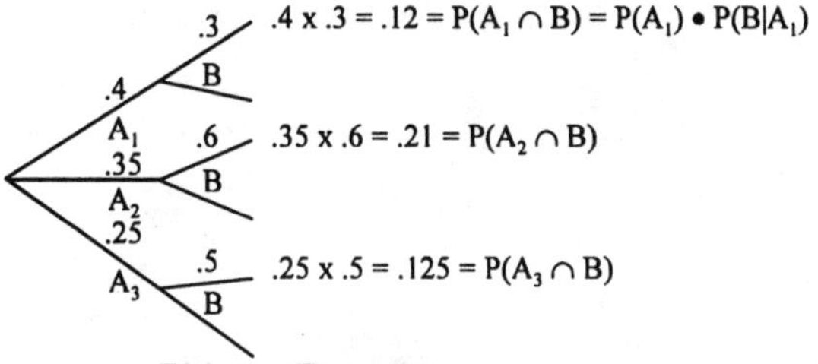

.4 — A_1 — .3 — B: $.4 \times .3 = .12 = P(A_1 \cap B) = P(A_1) \bullet P(B|A_1)$

.35 — A_2 — .6 — B: $.35 \times .6 = .21 = P(A_2 \cap B)$

.25 — A_3 — .5 — B: $.25 \times .5 = .125 = P(A_3 \cap B)$

a $P(A_2 \cap B) = .21$

b $P(B) = P(A_1 \cap B) + P(A_2 \cap B) + P(A_3 \cap B) = .455$

c $$P(A_1|B) = \frac{P(A_1 \cap B)}{P(B)} = \frac{.12}{.455} = .264$$
$$P(A_2|B) = \frac{.21}{.455} = .462,\ P(A_3|B) = 1 - .264 - .462 = .274$$

61 $P(\text{0 def in sample} \mid \text{0 def in batch}) = 1$

$$P(\text{0 def in sample} \mid \text{1 def in batch}) = \frac{\binom{9}{2}}{\binom{10}{2}} = .800$$

Chapter 2

$$P(\text{1 def in sample} \mid \text{1 def in batch}) = \frac{\binom{9}{1}}{\binom{10}{2}} = .200$$

$$P(\text{0 def in sample} \mid \text{2 def in batch}) = \frac{\binom{8}{2}}{\binom{10}{2}} = .622$$

$$P(\text{1 def in sample} \mid \text{2 def in batch}) = \frac{\binom{2}{1}\binom{8}{1}}{\binom{10}{2}} = .356$$

$$P(\text{2 def in sample} \mid \text{2 def in batch}) = \frac{1}{\binom{10}{2}} = .022$$

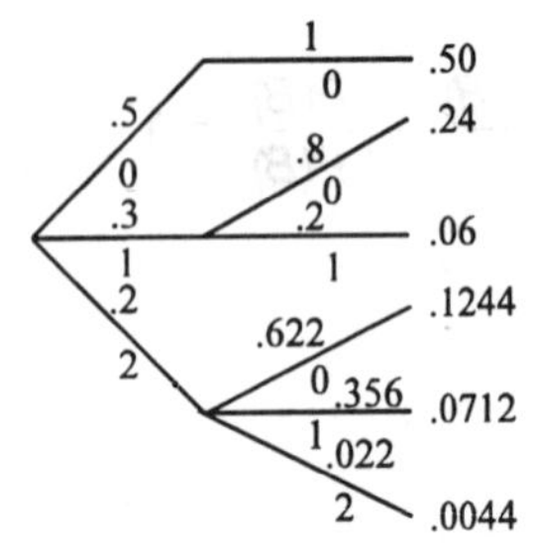

a $$P(\text{0 def in batch} \mid \text{0 def in sample}) = \frac{.5}{.5+.24+.1244} = .578$$

$$P(\text{1 def in batch} \mid \text{0 def in sample}) = \frac{.24}{.8644} = .278$$

$$P(\text{2 def in batch} \mid \text{0 def in sample}) = \frac{.1244}{.8644} = .144$$

b $$P(\text{0 def in batch} \mid \text{1 def in sample}) = 0$$

$$P(\text{1 def in batch} \mid \text{1 def in sample}) = \frac{.06}{.1312} = .457$$

$$P(\text{2 def in batch} \mid \text{1 def in sample}) = \frac{.0712}{.1312} = .543$$

63 **a**

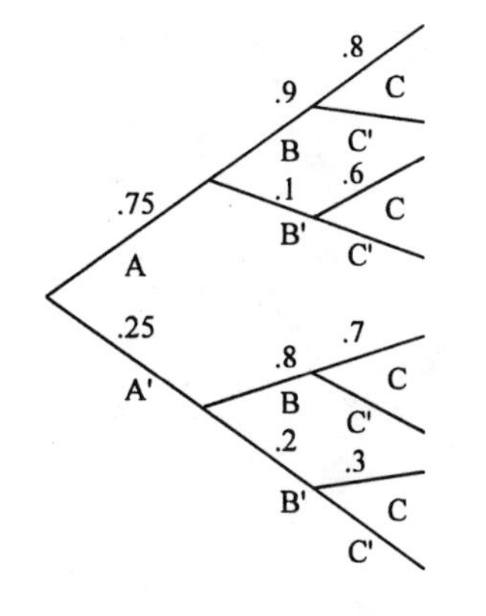

b $P(A \cap B \cap C) = .75 \times .9 \times .8 = .5400$

c $P(B \cap C) = P(A \cap B \cap C) + P(A' \cap B \cap C)$

$= .5400 + .25 \times .8 \times .7 = .6800$

d $P(C) = P(A \cap B \cap C) + P(A \cap B' \cap C)$

$+ P(A' \cap B \cap C) + P(A' \cap B' \cap C)$

$= .54 + .045 + .14 + .015 = .74$

e $P(A|B \cap C) = \dfrac{P(A \cap B \cap C)}{P(B \cap C)} = \dfrac{.54}{.68} = .7941$

65

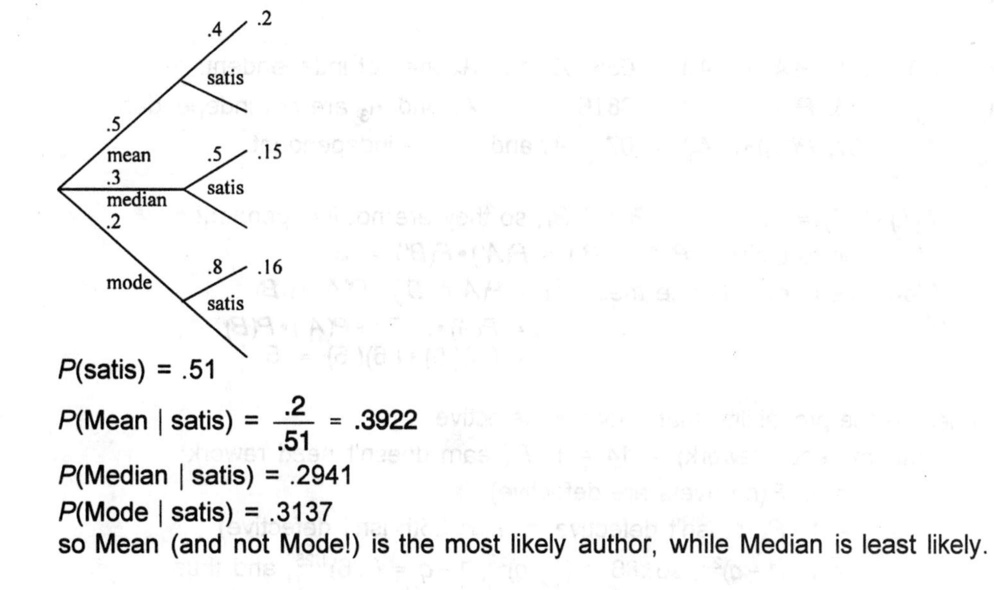

P(satis) = .51

P(Mean | satis) = $\dfrac{.2}{.51}$ = .3922

P(Median | satis) = .2941

P(Mode | satis) = .3137

so Mean (and not Mode!) is the most likely author, while Median is least likely.

67

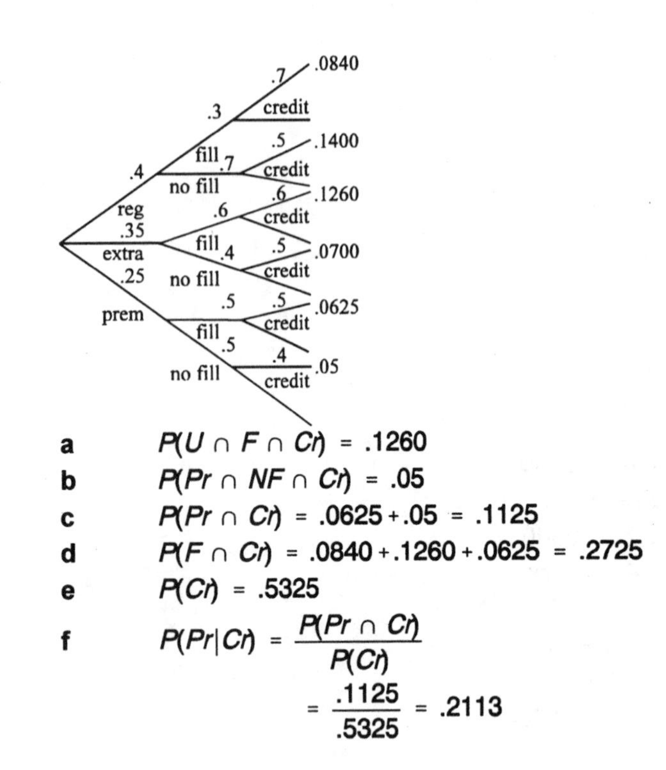

a $P(U \cap F \cap Cr) = .1260$

b $P(Pr \cap NF \cap Cr) = .05$

c $P(Pr \cap Cr) = .0625 + .05 = .1125$

d $P(F \cap Cr) = .0840 + .1260 + .0625 = .2725$

e $P(Cr) = .5325$

f $P(Pr|Cr) = \dfrac{P(Pr \cap Cr)}{P(Cr)}$

$= \dfrac{.1125}{.5325} = .2113$

Section 2.5

69 $P(A_1 \cap A_2) = .11$, $P(A_1) \bullet P(A_2) = .055$ A_1 and A_2 are not independent.

$P(A_1 \cap A_3) = .05$, $P(A_1) \bullet P(A_3) = .0616$ A_1 and A_3 are not independent.

$P(A_2 \cap A_3) = .07$, $P(A_2) \bullet P(A_3) = .07$ A_2 and A_3 *are* independent.

71 a $P(A) \bullet P(B) = .2 \neq .25 = P(A \cap B)$, so they are not independent.

b $P(\text{on time to both}) = P(A' \cap B') = P(A') \bullet P(B') = .3$

$P(\text{on time to exactly one meeting}) = P(A \cap B') + P(A' \cap B)$

$= P(A) \bullet P(B') + P(A') \bullet P(B)$

$= (.4)(.5) + (.6)(.5) = .5$

73 Let q denote the probability that a rivet is defective.

a $P(\text{seam needs rework}) = .14 = 1 - P(\text{seam doesn't need rework})$

$= 1 - P(\text{no rivets are defective})$

$= 1 - P(\text{1st isn't defective} \cap \ldots \cap \text{25th isn't defective})$

$= 1 - (1 - q)^{25}$, so $.86 = (1 - q)^{25}$, $1 - q = (.86)^{1/25}$, and thus

$q = 1 - .99399 = .00601$

b The desired condition is $.10 = 1-(1-q)^{25}$, i.e. $(1-q)^{25} = .90$, from which $q = 1-.99579 = .00421$.

75 Let A_1 = older pump fails, A_2 = newer pump fails, and $x = P(A_1 \cap A_2)$. Then $P(A_1) = .10+x$, $P(A_2) = .05+x$, and $x = P(A_1 \cap A_2) = P(A_1)\bullet P(A_2) = (.10+x)(.05+x)$. The resulting quadratic equation, $x^2-.85x+.005 = 0$, has roots $x = .0059$ and $x = .8441$. Hopefully the smaller root is the actual probability of system failure.

77 P(both detect the defect) = $1 - P$(at least one doesn't)

$$= 1-.2 = .8$$

a P(1st detects $\cap$ 2nd doesn't) = P(1st detects) $-$ P(1st does $\cap$ 2nd does)

$$= .9-.8 = .1$$

Similarly, P(1st doesn't $\cap$ 2nd does) = .1,

so P(exactly 1 does) = $.1+.1 = .2$.

b P(neither detects a defect) = $1-[P$(both do) $+ P$(exactly 1 does)]

$$= 1-[.8+.2] = 0,$$

so P(all 3 escape) = (0)(0)(0) = 0.

79 **a** Let D_1 = detection on 1st fixation, D_2 = detection on 2nd fixation.

P(detection in at most 2 fixations) $= P(D_1) + P(D_1' \cap D_2)$

$$= P(D_1) + P(D_2 | D_1')\bullet(P(D_1')$$

$$= p + p(1-p) = p(2-p)$$

b Define $D_1, D_2, \ldots, D_n$ as in **a**. Then P(at most n fixations)

$$= P(D_1) + P(D_1' \cap D_2) + P(D_1' \cap D_2' \cap D_3) + \ldots + P(D_1' \cap D_2' \cap \ldots \cap D_{n-1}' \cap D_n)$$

$$= p + p(1-p) + p(1-p)^2 + \ldots + p(1-p)^{n-1}$$

$$= p[1 + (1-p) + (1-p)^2 + \ldots + (1-p)^{n-1}] = p\bullet\frac{1-(1-p)^n}{1-(1-p)} = 1-(1-p)^n$$

Alternatively, P(at most n fixations) $= 1 - P$(at least $n+1$ are req'd)

$= 1 - P$(no detection in 1st n fixations)

$$= 1 - P(D_1' \cap D_2' \cap \ldots \cap D_n')$$

$$= 1-(1-p)^n$$

c P(no detection in 3 fixations) $= (1-p)^3$

d P(passes inspection) = P({not flawed} $\cup$ {flawed and passes})

$= P$(not flawed) + P(flawed and passes)

$= .9 + P$(passes | flawed) $\bullet$ P(flawed) $= .9 + (1-p)^3(.1)$

e $P(\text{flawed} \mid \text{passed}) = \dfrac{P(\text{flawed} \cap \text{passed})}{P(\text{passed})} = \dfrac{.1(1-p)^3}{.9+.1(1-p)^3}$

For $p = .5$, $P(\text{flawed} \mid \text{passed}) = \dfrac{.1(.5)^3}{.9+.1(.5)^3} = .0137$

81 $P(\text{system works}) = P(1-2 \text{ works} \cap 3-4-5-6 \text{ works} \cap 7 \text{ works})$
$= P(1-2 \text{ works}) \bullet P(3-4-5-6 \text{ works}) \bullet P(7 \text{ works})$
$= (.99)(.9639)(.9) = .8588$

With the subsystem in Example 2.36 connected in parallel to this subsystem,
$P(\text{system works}) = .8588 + .9639 - (.8588)(.9639) = .9949$

83

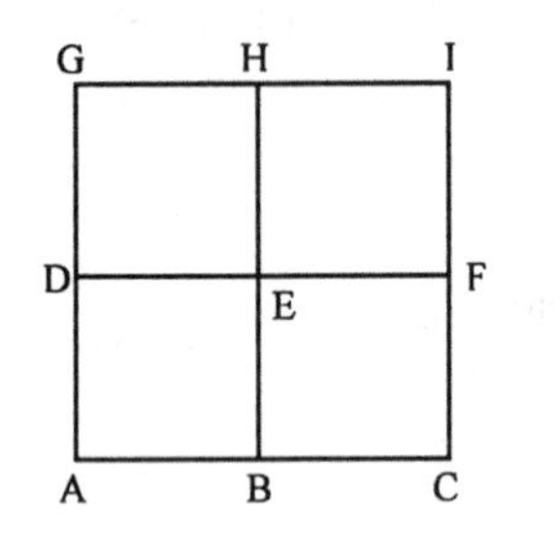

$P(\text{meet at } G) = P(A-D-G \cap I-H-G)$
$= P(A-D-G) \bullet P(I-H-G)$
$= \left(\frac{1}{2}\bullet\frac{1}{2}\right)\bullet\left(\frac{1}{2}\bullet\frac{1}{2}\right) = \frac{1}{16} = .0625$

Similarly, $P(\text{meet at } C) = \frac{1}{16}$ and

$P(\text{meet at } E) = P(A-D-E \text{ or } A-B-E) \bullet P(I-H-E \text{ or } I-F-E) = \frac{1}{2}\bullet\frac{1}{2} = \frac{1}{4}$, so

$P(\text{they do not meet}) = 1 - P(\text{meet}) = 1 - \frac{1}{4} - \frac{1}{16} - \frac{1}{16} = \frac{5}{8} = .625.$

Supplementary

85 **a** $\binom{20}{3} = 1140$ **b** $\binom{19}{3} = 969$

c # having at least 1 of the 10 best
= 1140 - # of crews having none of the 10 best
$= 1140 - \binom{10}{3} = 1140 - 120 = 1020$

d $P(\text{best will not work}) = \dfrac{969}{1140} = .85$

Chapter 2

87 All selection of six claims are equally likely.

$$P(\text{all disability examined}) = \frac{\binom{4}{4}\binom{6}{2}}{\binom{10}{6}} = \frac{15}{210} = .0714$$

P(only one of the two types remain) = P(all disab. examined) + P(all old age exam.)

$$= .0714 + = \frac{\binom{4}{0}\binom{6}{6}}{\binom{10}{6}} = .0714 + \frac{1}{210} = .0762$$

89 $P(A) \bullet P(B) = .0002,\ P(A \cup B) = P(A) + P(B) - P(A \cap B) = P(A) + P(B) - .0002 = .03$
$\Rightarrow P(B) = .0302 - P(A) \Rightarrow P(A)[.0302 - P(A)] = .0002$
$\Rightarrow [P(A)]^2 - .0302P(A) + .0002 = 0$, i.e. $x^2 - .0302x + .0002 = 0$
Solving yields $x \approx \dfrac{.0302 \pm .0106}{2} \Rightarrow P(A) \approx .0204,\ P(B) \approx .0098.$

91 **a** There are $5 \times 4 \times 3 \times 2 \times 1 = 120$ possible orderings, so

$$P(BCDEF) = \frac{1}{120} = .0083.$$

b # of orderings in which F is third = $4 \times 3 \times 1 \times 2 \times 1 = 24$, so (the 1 is "for F")

$$P(F\text{ 3rd}) = \frac{24}{120} = .2$$

c $$P(F\text{ last}) = \frac{4 \times 3 \times 2 \times 1 \times 1}{120} = .2$$

93

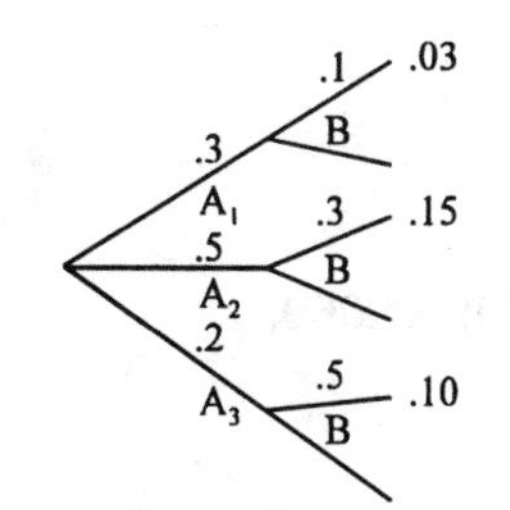

where B = at least 1 citation
P(at least 1 citation) = $.03 + .15 + .10 = .28$

$$P(\text{good} \mid \text{at least 1 citation}) = \frac{.03}{.28} = .1071$$

$$P(\text{medium} \mid \text{at least 1 citation}) = \frac{.15}{.28} = .5357$$

95 When three experiments are performed, there are three different ways in which detection can occur on exactly two of the experiments: (i) #1 and #2 and not #3; (ii) #1 and not #2 and #3; (iii) not #1 and #2 and #3. If the impurity is present, the probability of exactly two detections in three (independent) experiments is

$$(.8)(.8)(.2) + (.8)(.2)(.8) + (.2)(.8)(.8) = .384.$$

If the impurity is absent, the analogous probability is $3(.1)(.1)(.9) = .027$. Thus

$$P(\text{present} \mid \text{detected in exactly 2 out of 3}) = \frac{P(\text{detected in exactly 2} \cap \text{present})}{P(\text{detected in exactly 2})}$$

$$= \frac{(.384)(.4)}{(.384)(.4) + (.027)(.6)} = .905$$

97 **a**

$$P(\text{both +ve}) = P(\text{carrier} \cap \text{both + ve}) + P(\text{not a carrier} \cap \text{both +ve})$$

$$= P(\text{both +ve} \mid \text{carrier}) \bullet P(\text{carrier}) + P(\text{both +ve} \mid \text{not a carrier}) \bullet P(\text{not a carrier})$$

$$= (.90)^2(.01) + (.05)^2(.99) = .01058$$

$$P(\text{both} - \text{ve}) = (.10)^2(.01) + (.95)^2(.99) = .89358$$

$$P(\text{tests agree}) = .01058 + .89358 = .90416$$

b

$$P(\text{carrier} \mid \text{both +ve}) = \frac{P(\text{carrier} \cap \text{both positive})}{P(\text{both +ve})}$$

$$= \frac{(.90)^2 \bullet (.01)}{.01058} = .7656$$

99 $P(E_1 \cap \text{late}) = P(\text{late} \mid E_1) \bullet P(E_1) = (.02)(.40) = .008$

101 Let B denote the event that a component needs rework. Then

$$P(B) = \sum_{i=1}^{3} P(B \mid A_i) \bullet P(A_i) = (.05)(.50) + (.08)(.30) + (.10)(.20) = .069.$$

Thus $P(A_1 \mid B) = \dfrac{(.05)(.50)}{.069} = .362$

$$P(A_2 \mid B) = \frac{(.08)(.30)}{.069} = .348$$

$$P(A_3 \mid B) = \frac{(.10)(.20)}{.069} = .290$$

Chapter 2

103

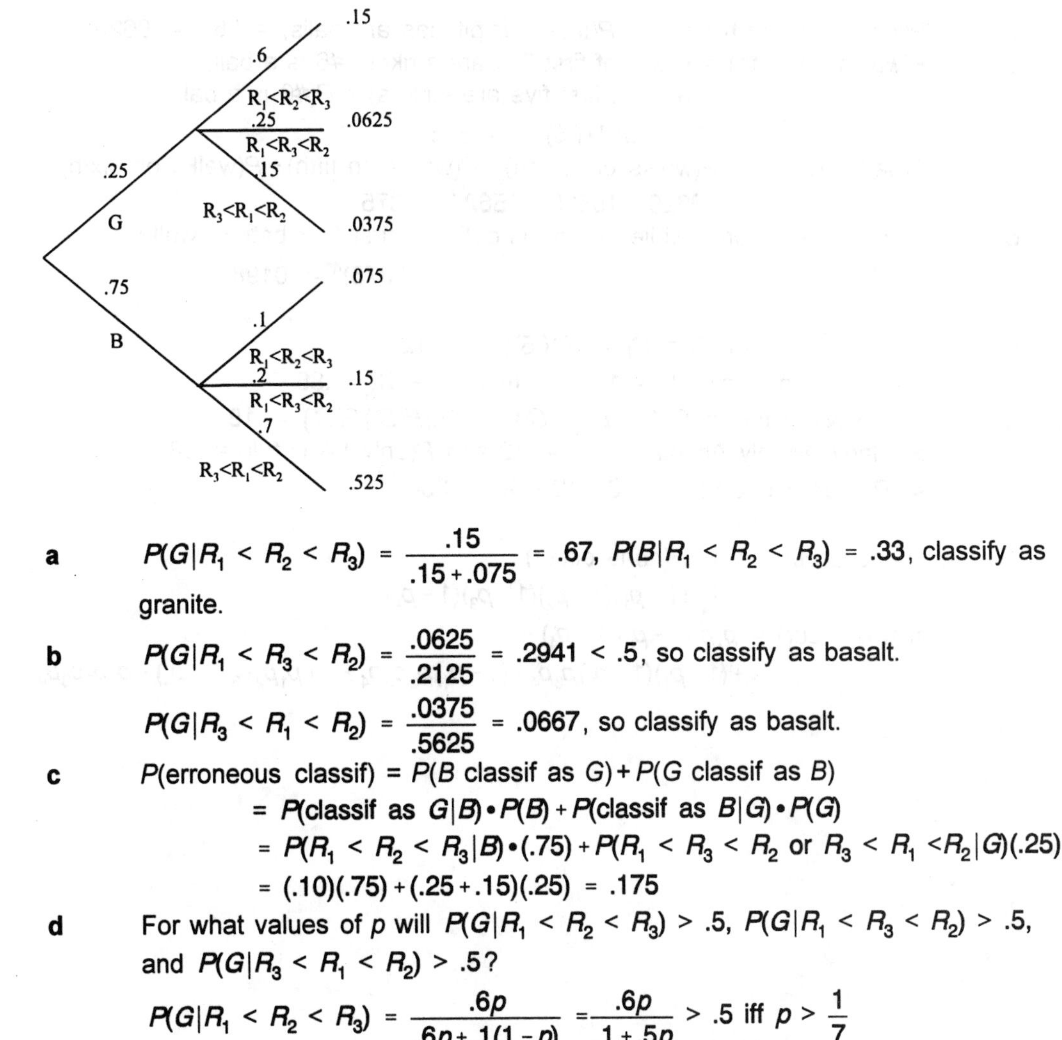

a $P(G|R_1 < R_2 < R_3) = \dfrac{.15}{.15+.075} = .67$, $P(B|R_1 < R_2 < R_3) = .33$, classify as granite.

b $P(G|R_1 < R_3 < R_2) = \dfrac{.0625}{.2125} = .2941 < .5$, so classify as basalt.

$P(G|R_3 < R_1 < R_2) = \dfrac{.0375}{.5625} = .0667$, so classify as basalt.

c $P(\text{erroneous classif}) = P(B \text{ classif as } G) + P(G \text{ classif as } B)$

$= P(\text{classif as } G|B) \bullet P(B) + P(\text{classif as } B|G) \bullet P(G)$

$= P(R_1 < R_2 < R_3|B) \bullet (.75) + P(R_1 < R_3 < R_2 \text{ or } R_3 < R_1 < R_2|G)(.25)$

$= (.10)(.75) + (.25+.15)(.25) = .175$

d For what values of p will $P(G|R_1 < R_2 < R_3) > .5$, $P(G|R_1 < R_3 < R_2) > .5$, and $P(G|R_3 < R_1 < R_2) > .5$?

$P(G|R_1 < R_2 < R_3) = \dfrac{.6p}{.6p+.1(1-p)} = \dfrac{.6p}{.1+.5p} > .5$ iff $p > \dfrac{1}{7}$

$P(G|R_1 < R_3 < R_2) = \dfrac{.25p}{.25p+.2(1-p)} > .5$ iff $p > \dfrac{4}{9}$

$P(G|R_3 < R_1 < R_2) = \dfrac{.15p}{.15p+.7(1-p)} > .5$ iff $p > \dfrac{14}{17} \leftarrow$ most restrictive

If $p > \dfrac{14}{17}$ always classify as granite

Chapter 2

105 **a** P(walks on fourth pitch) = P(first four pitches are balls) = $(.5)^4 = .0625$

b P(walks on sixth) = P(two of first five are strikes, #6 is a ball)

= P(two of first five are strikes) • P(#6 is a ball)

$= [10 \bullet (.5)^5] \bullet (.5) = .15625$

c P(batter walks) = P(walks on fourth) + P(walks on fifth) + P(walks on sixth)

$= .0625 + .15625 + .15625 = .375$

d P(first batter scores while no one is out) = P(first four batters walk)

$= (.375)^4 = .0198$

107 **a** $P(\text{all full}) = P(A \cap B \cap C) = (.6)(.5)(.4) = .12$

$P(\text{at least one isn't full}) = 1 - P(\text{all full}) = 1 - .12 = .88$

b $P(\text{only NY is full}) = P(A \cap B' \cap C') = P(A)P(B')P(C') = .18$

Similarly, P(only Atlanta is full) = .12 and P(only LA is full) = .08,

so $P(\text{exactly one full}) = .18 + .12 + .08 = .38$.

109 $P(\text{at least one occurs}) = 1 - P(\text{none occur})$

$= 1 - (1 - p_1)(1 - p_2)(1 - p_3)(1 - p_4)$

$P(\text{at least two occur}) = p_1 p_2 (1 - p_3)(1 - p_4) + \ldots$

$+ (1 - p_1)(1 - p_2) p_3 p_4 + (1 - p_1) p_2 p_3 p_4 + \ldots + p_1 p_2 p_3 (1 - p_4) + p_1 p_2 p_3 p_4$

CHAPTER 3

Section 3.1

1

S:	FFF	SFF	FSF	FFS	FSS	SFS	SSF	SSS
X:	0	1	1	1	2	2	2	3

3 M = the difference between the larger and the smaller outcome with possible values 0, 1, 2, 3, 4, or 5; $W = 1$ if the sum of the two resulting numbers is even and $W = 0$ otherwise, a Bernoulli random variable

5 No. In the experiment in which a coin is tossed repeatedly until a H results, let $Y = 1$ if the experiment terminates with at most 5 tosses and $Y = 0$ otherwise. The sample space is infinite, yet Y has only two possible values.

7
- **a** Possible values are 0, 1, 2, ..., 12; discrete
- **b** With N = # on the list, values are 0, 1, 2, ..., N; discrete
- **c** Possible values are 1, 2, 3, 4, ...; discrete
- **d** $\{ x: 0 < x < \infty \}$ if we assume that a rattlesnake can be arbitrarily short or long; not discrete
- **e** With c = amount earned per book sold, possible values are 0, c, $2c$, $3c$, ..., $10{,}000c$; discrete
- **f** $\{ y: 0 < y < 14 \}$ since 0 is the smallest possible pH and 14 is the largest possible pH; not discrete
- **g** With m and M denoting the minimum and maximum possible tension, respectively, possible values are $\{ x: m \le x \le M \}$; not discrete
- **h** Possible values are 3, 6, 9, 12, 15,... - i.e. 3(1), 3(2), 3(3), 3(4),...giving a first element, etc. ; discrete

9
- **a** Returns to 0 can occur only after an even number of tosses; possible X values are 2, 4, 6, 8, ... (i.e. 2(1), 2(2), 2(3), 2(4), ...), an infinite sequence, so X is discrete
- **b** Now a return to 0 is possible after any number of tosses greater than 1, so possible values are 2, 3, 4, 5, ...
  $(1+1,\ 1+2,\ 1+3,\ 1+4, \ldots,$ an infinite sequence$)$, and X is discrete

Section 3.2

11 **a**

x	4	6	8
$p(x)$	.45	.40	.15

b

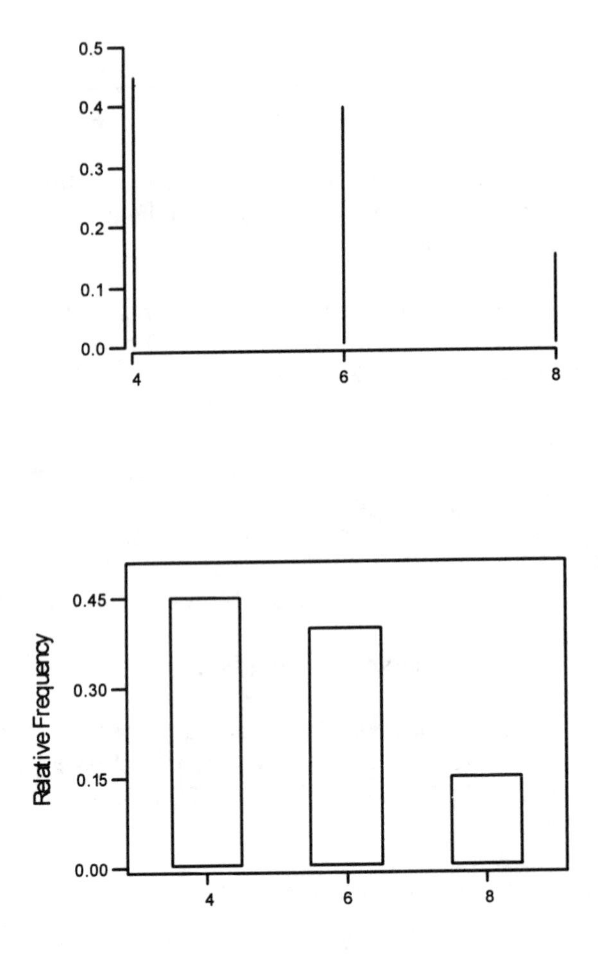

13 **a** $P(X \le 3) = p(0) + p(1) + p(2) + p(3) = .10 + .15 + .20 + .25 = .70$

b $P(X < 3) = P(X \le 2) = p(0) + p(1) + p(2) = .45$

c $P(3 \le X) = p(3) + p(4) + p(5) + p(6) = .55$

d $P(2 \le X \le 5) = p(2) + p(3) + p(4) + p(5) = .71$

e The number of lines not in use is $6 - X$, so $6 - X = 2$ is equivalent to $X = 4$, $6 - X = 3$ to $X = 3$, and $6 - X = 4$ to $X = 2$. Thus we desire $P(2 \le X \le 4) = p(2) + p(3) + p(4) = .65$

f $6 - X \ge 4$ if $6 - 4 \ge X$, i.e. $2 \ge X$, i.e. $X \le 2$ and $P(X \le 2) = .10 + .15 + .20 = .45$

Chapter 3

15 **a** (1, 2), (1, 3), (1, 4),(1, 5), (2, 3), (2, 4), (2, 5), (3, 4), (3, 5), (4, 5)

b $P(X = 0) = p(0) = P(\{(3,4),(3,5),(4,5)\}) = \dfrac{3}{10} = .30$

$P(X = 2) = p(2) = P(\{1,2\}) = \dfrac{1}{10} = .10$

$P(X = 1) = p(1) = 1 - [p(0) + p(2)] = .60$

and $p(x) = 0$ if $x \neq 0, 1,$ or 2

c $F(0) = P(X \leq 0) = P(X = 0) = .30$

$F(1) = P(X \leq 1) = P(X = 0 \text{ or } 1) = .90$

$F(2) = P(X \leq 2) = 1$

The c.d.f. is thus

$$F(x) = \begin{cases} 0 & x < 0 \\ .30 & 0 \leq x < 2 \\ .90 & 1 \leq x < 2 \\ 1 & 2 \leq x \end{cases}$$

17 **a** $p(2) = P(Y = 2) = P$ (first 2 batteries are acceptable)

$= P(AA) = (.8)(.8) = .64$

b $p(3) = P(Y = 3) = P(UAA \text{ or } AUA) = (.2)(.8)^2 + (.2)(.8)^2$

$= 2(.2)(.8)^2 = .256$

c The fifth battery must be an A, and one of the first four must also be an A. Thus

$p(5) = P(AUUUA \text{ or } UAUUA \text{ or } UUAUA \text{ or } UUUAA)$

$= 4(.2)^3(.8)^2$ or $.02048$

d $P(Y = y) = p(y) = P$ (the yth is an A and so is exactly one of the first $y - 1) = (y-1)(.2)^{y-2}(.8)^2 \qquad y = 2, 3, 4, 5, \ldots$

19 Let A denote the type 0+ individual (type 0 positive blood) and B, C, D the other three individuals. Then

$p(1) = P(Y = 1) = P(A \text{ first}) = \dfrac{1}{4} = .25$

$p(2) = P(Y = 2) = P(B, C, \text{ or } D \text{ first and } A \text{ next}) = \dfrac{3}{4} \cdot \dfrac{1}{3} = \dfrac{1}{4} = .25$

$p(4) = P(Y = 4) = P(A \text{ last}) = \dfrac{3}{4} \cdot \dfrac{2}{3} \cdot \dfrac{1}{2} = \dfrac{1}{4} = .25$

so $p(3) = 1 - [.25 + .25 + .25] = .25$

21 The jumps in $F(x)$ occur at $x = 0, 1, 2, 3, 4, 5,$ and 6, so we first calculate $F(\cdot)$ at each of these values:

$F(0) = P(X \le 0) = P(X = 0) = .10$
$F(1) = P(X \le 1) = p(0) + p(1) = .25$
$F(2) = P(X \le 2) = p(0) + p(1) + p(2) = .45$
$F(3) = .70, F(4) = .90, F(5) = .96,$ and $F(6) = 1.$
Thus

$$F(x) = \begin{cases} 0 & x < 0 \\ .10 & 0 \le x < 1 \\ .25 & 1 \le x < 2 \\ .45 & 2 \le x < 3 \\ .70 & 3 \le x < 4 \\ .90 & 4 \le x < 5 \\ .96 & 5 \le x < 6 \\ 1.00 & 6 \le x \end{cases}$$

Then $P(X \le 3) = F(3) = .70, P(X < 3) = P(X \le 2) = F(2) = .45,$
$P(3 \le X) = 1 - P(X \le 2) = 1 - F(2) = 1 - .45 = .55,$ and
$P(2 \le X \le 5) = F(5) - F(1) = .96 - .25 = .71$

23 **a** Possible X values are those values at which $F(x)$ jumps, and the probability of any particular value is the size of the jump at that value. Thus we have

x	1	3	4	6	12
$p(x)$	.30	.10	.05	.15	.40

b $P(3 \le X \le 6) = F(6) - F(3-) = .60 - .30 = .30$
$P(4 \le X) = 1 - P(X < 4) = 1 - F(4-) = 1 - .40 = .60$

25 **a** Possible X values are 1, 2, 3, ...

$p(1) = P(X = 1) = P$ (return home after just one visit) $= \frac{1}{3}$

$p(2) = P(X = 2) = P$ (second visit and then return home) $\frac{2}{3} \cdot \frac{1}{3}$

$p(3) = P(X = 3) = P$ (three visits and then home) $= \left(\frac{2}{3}\right)^2 \cdot \left(\frac{1}{3}\right)$

In general, $p(x) = \left(\frac{2}{3}\right)^{x-1}\left(\frac{1}{3}\right)$ for $x = 1,2,3,...$

b The number of straight line segments is $Y = 1 + X$ (since the last segment traversed returns Alvie to 0), so as in (a), $p(y) = \left(\frac{2}{3}\right)^{y-2}\left(\frac{1}{3}\right)$ for $y = 2,3,\ldots$

c Possible Z values are 0, 1, 2, 3, ... $p(0) = P$ (male first and then home) $= \left(\frac{1}{2}\right)\left(\frac{1}{3}\right) = \frac{1}{6}$, $p(1) = P$ (exactly one visit to a female) $= P$ (female first, then home) $+ P$ (female first, then male, then home) $+ P$ (male first, then female, then home) $+ P$ (male first, then female, then male, then home)

$$= \left(\frac{1}{2}\right)\left(\frac{1}{3}\right) + \left(\frac{1}{2}\right)\left(\frac{2}{3}\right)\left(\frac{1}{3}\right) + \left(\frac{1}{2}\right)\left(\frac{2}{3}\right)\left(\frac{1}{3}\right) + \left(\frac{1}{2}\right)\left(\frac{2}{3}\right)\left(\frac{2}{3}\right)\left(\frac{1}{3}\right) = \left(\frac{1}{2}\right)\left(1 + \frac{2}{3}\right)\left(\frac{1}{3}\right)$$

$$+ \left(\frac{1}{2}\right)\left(\frac{2}{3}\right)\left(\frac{2}{3} + 1\right)\left(\frac{1}{3}\right) = \left(\frac{1}{2}\right)\left(\frac{5}{3}\right)\left(\frac{1}{3}\right) + \left(\frac{1}{2}\right)\left(\frac{2}{3}\right)\left(\frac{5}{3}\right)\left(\frac{1}{3}\right)$$

where the first term corresponds to initially visiting a female and the second term corresponds to initially visiting a male. Similarly,

$p(2) = \left(\frac{1}{2}\right)\left(\frac{2}{3}\right)^2\left(\frac{5}{3}\right)\left(\frac{1}{3}\right) + \left(\frac{1}{2}\right)\left(\frac{2}{3}\right)^2\left(\frac{5}{3}\right)\left(\frac{1}{3}\right)$. In general,

$p(z) = \left(\frac{1}{2}\right)\left(\frac{2}{3}\right)^{2z-2}\left(\frac{5}{3}\right)\left(\frac{1}{3}\right) + \left(\frac{1}{2}\right)\left(\frac{2}{3}\right)^{2z-2}\left(\frac{5}{3}\right)\left(\frac{1}{3}\right) = \left(\frac{25}{54}\right)\left(\frac{2}{3}\right)^{2z-2}$ for $z = 1,2,3,\ldots$

27 If $x_1 < x_2$, $F(x_2) = P(X \le x_2) = P(\{X \le x_1\} \cup \{x_1 < X \le x_2\})$
$= P(X \le x_1) + P(x_1 < X \le x_2) \ge P(X \le x_1) = F(x_1)$.
Equality obtains if $P(x_1 < X \le x_2) = 0$.

Section 3.3

29 **a** $E(Y) = \sum_{y=1}^{4} y \cdot p(y) = (1)(.01) + (2)(.19) + (3)(.35) + (4)(.45) = 3.24$

b $E(\sqrt{Y}) = \sum_{y=1}^{4} \sqrt{y} \cdot p(y) = (\sqrt{1})(.01) + (\sqrt{2})(.19) + (\sqrt{3})(.35) + (\sqrt{4})(.45) = 1.785$

31 **a** $E(X) = (13.5)(.2) + (15.9)(.5) + (19.1)(.3) = 16.38$,
$E(X^2) = (13.5)^2(.2) + (15.9)^2(.5) + (19.1)^2(.3) = 272.298$,
$V(X) = 272.298 - (16.38)^2 = 3.9936$

b $E[25X - 8.5] = 25E(X) - 8.5 = (25)(16.38) - 8.5 = 401$

c $V[25X - 8.5] = V[25X] = (25)^2 V(X) = (625)(3.9936) = 2496$

d $E[h(X)] = E[X - .01X^2] = E(X) - .01E(X^2) = 16.38 - 2.72 = 13.66$

Chapter 3

33 $E(X) = \sum_{x=1}^{\infty} x \cdot p(x) = \sum_{x=1}^{\infty} x \cdot \frac{c}{x^3} = c \sum_{x=1}^{\infty} \frac{1}{x^2}$, but it is a well-known result from the theory of infinite series that $\sum_{x=1}^{\infty} \frac{1}{x^2} < \infty$, so $E(X)$ is finite

35

$p(x)$	.8	.1	.08	.02
x	0	1,000	5,000	10,000
$h(x)$	0	500	4,500	9,500

$E(h(x)) = 600$
premium should be \$100 plus expected value of damage minus deductible or \$700

37 $E(h(X)) = E\left(\frac{1}{X}\right) = \sum_{x=1}^{6} \left(\frac{1}{x}\right) \cdot p(x) = \frac{1}{6} \sum_{x=1}^{6} \frac{1}{x} = .408$, whereas $\frac{1}{3.5} = .286$, so you expect to win more if you gamble.

39 **a** The line graph of the pmf of $-X$ is just the line graph of the p.m.f. of X reflected about zero, (the pmf of $-X$ puts probability $\frac{1}{6}$ on the values $-1, -2, -3, -4, -5$, and -6), but both have the same degree of spread about their respective means, suggesting $V(-X) = V(X)$.

b With $a = -1, b = 0, V(aX+b) = V(-X) = a^2V(X) = V(X)$.

41 **a** $E[X(X-1)] = E(X^2) - E(X) \Rightarrow E(X^2) = E[X(X-1)] + E(X) = 32.5$

b $V(X) = 32.5 - (5)^2 = 7.5$

c $V(X) = E[X(X-1)] + E(X) - [E(X)]^2$

43 **a**

k	2	3	4	5	10
$\frac{1}{k^2}$	.25	.11	.06	.04	.01

b $\mu = \sum_{x=0}^{6} x \cdot p(x) = 2.64, \sigma^2 = \sum x^2 \cdot p(x) - \mu^2 = 2.37, \sigma = 1.54.$

Thus $\mu - 2\sigma = -.44$ and $\mu + 2\sigma = 5.72$, so $P(\mid X - \mu \mid \geq 2\sigma)$

$= P(X$ is at least 2 s.d.'s from $\boldsymbol{\mu})$

$= P(X$ is either $\leq -.44$ or $\geq 5.72) = P(X = 6) = .04$

Chebyshev's bound of 0.25 is much too conservative. For $k = 3,4,5,$ and $10, P(|X - \mu| \geq k\sigma) = 0$ here, again pointing to the very conservative nature of the bound $\frac{1}{k^2}$.

c $\mu = 0$ and $\sigma = \frac{1}{3}$, so $P(\mid X - \mu \mid \geq 3\sigma) = P(\mid X \mid \geq 1)$

$= P(X = -1 \text{ or } +1) = \frac{1}{18} + \frac{1}{18} = \frac{1}{9}$, identical to the upper bound.

d Let $p(-1) = \frac{1}{50}, p(+1) = \frac{1}{50}, p(0) = \frac{24}{25}$.

Section 3.4

45

a $B(4;10,.3) = .850$

b $b(4;10,.3) = B(4;10,.3) - B(3;10,.3) = .200$

c $b(6;10,.7) = B(6;10,.7) - B(5;10,.7) = .200$

d $P(2 \leq X \leq 4) = B(4;10,.3) - B(1;10,.3) = .701$

e $P(2 < X) = 1 - P(X \leq 1) = 1 - B(1;10,.3) = .851$

f $P(X \leq 1) = B(1;10,0.7) = 0.000$

g $P(2 < X < 6) = P(3 \leq X \leq 5) = B(5;10,.3) - B(2;10,.3) = .570$

47 $X \sim \text{Bin}(6, .10)$

a $P(X = 1) = \binom{n}{x}(p^x)(1-p)^{n-x}$

$= \binom{6}{1}(.1)^1(.9)^5$

$= .354$

b $P(X \geq 2) = 1 - [P(X = 0) + P(X = 1)]$ we know that $P(X = 1) = .3543$

$P(X = 0) = \binom{6}{0}(.1)^0(.9)^6 = .5314$ hence

$P(X \geq 2) = 1 - [.3543 + .5314] = .115$

c Either four or five goblets must be selected

i. select four goblets with zero defects $P(X = 0) = \binom{4}{0}(.1)^0(.9)^4 = .6561$

ii. select four goblets with one defect and the fifth goblet is good

$\left[\binom{4}{1}(.1)^1(.9)^3\right] \times .9 = .26244$

$.6561 + .26244 = .918$

49 Let S = fail to run on the first submission, so $p = .9, n = 15$

a $P(12 \le x) = 1 - P(X \le 11) = 1 - B(11;15,.9) = .944$

b $P(10 \le X \le 13) = B(13;15,.9) - B(9;15,.9) = .449$

c P (at most 2 F's) = P (at least 13 S's) $= P(13 \le X) = 1 - P(X \le 12)$

$= 1 - B(12;15,.9) = .816$ (equivalently, $B(2;15,.1)$).

51 Let S represent a telephone that is submitted for service while under warranty and must be replaced. Then $p = P(S) = P$ (replaced | submitted) $\cdot P$ (submitted) $= (.40)(.20) = .08$. Thus X, the number among the company's ten phones that must be replaced, has a binomial distribution with

$n = 10, p = .08$, so $p(2) = P(X = 2) = \binom{10}{2}(.08)^2(.92)^8 = .1478$

53 **a** P (rejecting claim when $p = .8) = B(15;25,.8) = .017$

b P (not rejecting claim when $p = .7) = P(X \ge 16$ when $p = .7)$

$= 1 - B(15;25,.7) = 1 - .189 = .811$; for $p = .6$,

this probability is $1 - B(15;25,.6) = 1 - .575 = .425$

c The probability of rejecting the claim when $p = .8$ becomes $B(14;25,.8) = .006$, smaller than in (a) above. However, the probabilities of (b) above increase to .902 and .586, respectively.

55 If topic A is chosen, P (at least half received)

$= P(1 \le X)$ (when $n = 2) = 1 - P(X = 0) = 1 - (.1)^2 = .99$.

If B is chosen, P (at least half received)

$= P(2 \le X)$ (when $n = 4) = 1 - P(X \le 1) = 1 - (0.1)^4 - 4(.1)^3(.9) = .9963$.

Thus topic B should be chosen. If $p = .5$, the probabilities are .75 for A and .6875 for B, so now A should be chosen.

57 **a** $b(x;n,1-p) = \binom{n}{x}(1-p)^x p^{n-x} = \binom{n}{n-x} p^{n-x}(1-p)^x = b(n-x;n,p)$.

Alternatively, $P(x S' s$ when $P(S) = 1-p) = P(n-x F' s$ when $P(F) = p)$ (since the two events are identical), but the labels S and F are arbitrary so can be interchanged (if $P(S)$ and $P(F)$ are also interchanged), yielding $P(n-x S's \text{ when} P(S) = 1-p)$ as desired.

b $B(x;n,1-p) = P$ (at most x $S's$ when $P(S) = 1-p) = P$ (at least $n-x$ $F's$ when $P(F) = P$ (at least $n-x$ $S's$ when $P(S) = P) = 1-P$ (at most $n-x-1$ $S's$ when $P(S) = p) = 1-B(n-x-1;n,p)$

c Whenever $p > .5, 1-p < .5$, so probabilities involving X can be calculated using the results (a) and (b) in combination with tables giving probabilities only for $p \le .5$.

59 **a** Let $X =$ the number who select leaded regular fuel. Then X is a binomial random variable with $n = 100$ and $p = .3$, so $E(X) = 30$ and $V(X) = 21$

b With $X =$ the number who select leaded fuel, $p = P(A) + P(B) = .5$ so $E(X) = 50$ and $V(X) = 25$.

61 When $p = .5, \mu = 10$ and $\sigma = 2.236$, so $2\sigma = 4.472$ and $3\sigma = 6.708$. The inequality $|X-10| \ge 4.472$ is satisfied if either $X \le 5$ or $X \ge 15$, or $P(|X-\mu| \ge 2\sigma) = P(X \le 5 \text{ or } \ge 15) = .021 + .021 = .042$. Similarly, $P(|X-\mu| \ge 3\sigma) = P(X \le 3 \text{ or } \ge 17) = .002$. In the case $p = .75, \mu = 15$ $\sigma = 1.937$, giving $2\sigma = 3.874$ and $3\sigma = 5.811$. Thus $P(|X-15| \ge 3.874) = P(X \le 11 \text{ or } \ge 19) = .041 + .024 = .065$, whereas $P(|X-15| \ge 5.811) = P(X \le 9) = .004$. All these probabilities are considerably less than the upper bounds $.25 (k = 2)$ and $.11 (k = 3)$ given by Chebyshev.

Section 3.5

63 **a** $$P(X = 1) = h(1;6,4,12) = \frac{\binom{4}{1}\binom{8}{5}}{\binom{12}{6}} = .242$$

b Since at most four can be defective, $P(X \ge 4) = P(X = 4) = h(4;6,4,12) = .030$

c $P(1 \le X \le 3) = 1 - P(X = 0) - P(X = 4) = 1 - .0303 - .0303 = .939$

65 With $X =$ the number of basalt specimens collected, $p(x) = h(x;15,10,20)$. P (all specimens of one of the two types are selected) $P(X = 10) + P(X = 5)$ (since if $X = 5$ all granite specimens are selected) $= h(10;15,10,20) + h(5;15,10,20) = .033$

67 **a** $h(x;10,10,20)$ (the successes here are the top 10 pairs, and a sample of 10 pairs is drawn from among the 20)

b Let $X =$ the number among the top five who play E-W. Then P (all of top five play the same direction)

$$= P(X = 5) + P(X = 0) = h(5;10,5,20) + h(0;10,5,20) = \frac{\binom{15}{5}}{\binom{20}{10}} + \frac{\binom{15}{10}}{\binom{20}{10}} = .033$$

c $N = 2n;\ M = n;\ n = n$

$h(x;\ n,\ n,\ 2n)$

$$E(X) = n \cdot \frac{n}{2n} = \frac{1}{2}n$$

$$V(X) = \left(\frac{2n-n}{2n-1}\right) \cdot n\frac{n}{2n} \cdot \left(1 - \frac{n}{2n}\right) = \left(\frac{n}{2n-1}\right) \cdot \frac{n}{2} \cdot \left(1 - \frac{n}{2n}\right) = \left(\frac{n}{2n-1}\right) \cdot \frac{n}{2} \cdot \left(\frac{1}{2}\right)$$

69 **a** With $S =$ a female child and $F =$ a male child, let $X =$ the number of F's before the 2nd S. Then $P(X = x) = nb(x;2,.5)$

b P (exactly four children) = P (exactly two males)

$$= nb(2;2,.5) = (3)(.0625) = .188$$

c P (at most four children)

$$= P(X \le 2) = \sum_{x=0}^{2} nb(x;2,.5) = .25 + 2(.25)(.5) + 3(.0625) = .688$$

d $E(X) = \frac{(2)(.5)}{(.5)} = 2$, expected number of children $= E(X+2) = E(X) + 2 = 4$

71 This is identical to an experiment in which a single family has children until exactly six females have been born (since $p = .5$ for each of the three families), so $p(x) = nb(x;6,.5)$ and $E(X) = 6$ ($= 2+2+2$, the sum of the expected number of males born to each one).

Section 3.6

73 **a** $P(X \le 8) = F(8;5) = .932$

b $P(X = 8) = F(8;5) - F(7;5) = .065$

c $P(9 \le X) = 1 - P(X \le 8) = .068$

d $P(5 \le X \le 8) = F(8;5) - F(4;5) = .492$

e $P(5 < X < 8) = .251$

75 **a** $P(X \le 10) = F(10;10) = .583$

b $P(10 \le X \le 15) = F(15;10) - F(9;10) = .493$

c $E(X) = \lambda = 10, \sigma_X = \sqrt{\lambda} = 3.16$

Chapter 3

77 $p = \frac{1}{200}$; $n = 1{,}000$

$\lambda = np = 5$

a $P(4 \le x \le 7) = P(x \le 7) - P(x \le 3)$
$= .867 - .265 = .602$

b $P(x \ge 8) = 1 - P(x \le 7) = 1 - .867 = .133$

79 **a** $\lambda = 8$ when $t = 1$, so $P(X = 5) = F(5;8) - F(4;8) = .191 - .100 = .091$,
$P(X \ge 5) = 1 - F(4;8) = .900$, and $P(X \ge 10) = 1 - F(9;8) = .283$

b $t = 90$ min $= 1.5$ hours, so $\lambda = 12$; thus the expected number of arrivals is 12 and the s.d. is $\sqrt{12} = 3.464$.

c $t = 2.5$ hours implies that $\lambda = 20$; in this case,
$P(x \ge 20) = 1 - F(19;20) = .530$ and $P(X \le 10) = F(10;20) = .011$.

81 **a** For a 2-hour period the parameter of the distribution is
$\lambda t = (4)(2) = 8$, so $P(X = 10) = F(10;8) - F(9;8) = .099$

b For a 30-minute period, $\lambda t = (4)(.5) = 2$, so $P(X = 0) = F(0;2) = .135$

c $E(X) = \lambda t = 2$

83 $\alpha = 1/\text{mean time between occurrences} = \frac{1}{.5} = 2$

a $\alpha t = (2)(2) = 4$

b $P(X > 5) = 1 - P(X \le 5) = 1 - .785 = .215$

c solve for t given $\alpha = 2$

$.1 = e^{-\alpha t}$

$\ln(.1) = -\alpha t$

$t = \frac{2.3026}{2} \approx 1.1513$ years

85 **a** For a one-quarter acre plot, the parameter is
$(80)(.25) = 20$, so $P(X \le 16) = F(16;20) = .221$

b The expected number of trees is $\lambda \cdot (\text{area}) = 80(85{,}000) = 6{,}800{,}000$

c The area of the circle is $\pi r^2 = .031416$ sq. miles or 20.106 acres. Thus X has a Poisson distribution with parameter 20.106

87 **a** No events in $(0, t+\Delta t)$ if and only if no events in $(0, t)$and no events in $(t, t+\Delta t)$. Thus $P_0(t+\Delta t) = P_0(t) \cdot P$ (no events in $(t, t+\Delta t)$)
$= P_0(t)[1 - \lambda \cdot \Delta t - o(\Delta t)]$

b $\dfrac{P_0(t+\Delta t)-P_0(t)}{\Delta t} = -\lambda P_0(t)\,\dfrac{\Delta' t}{\Delta' t} - P_0(t)\cdot\dfrac{o(\Delta t)}{\Delta t}$

As $\Delta t \to 0, \dfrac{o(\Delta t)}{\Delta t} \to 0$ so letting $\Delta t \to 0$ yields $P_0'(t) = -\lambda P_0(t)$

c $\dfrac{d}{dt}[e^{-\lambda t}] = -\lambda e^{-\lambda t} = -\lambda P_0(t)$ as desired

d $\dfrac{d}{dt}\left[\dfrac{e^{-\lambda t}(\lambda t)^k}{k!}\right] = \dfrac{-\lambda e^{-\lambda t}(\lambda t)^k}{k!} + \dfrac{k\lambda e^{-\lambda t}(\lambda t)^{k-1}}{k!}$

$= -\lambda\dfrac{e^{-\lambda t}(\lambda t)^k}{k!} + \lambda\dfrac{e^{-\lambda t}(\lambda t)^{k-1}}{(k-1)!} = -\lambda P_k(t) + \lambda P_{k-1}(t)$ as desired

Supplementary

89 a $p(1) = P$ (exactly one suit) $= P$ (all spades) $+ P$ (all hearts) $+ P$ (all diamonds) $+ P$ (all clubs) $= 4P$ (all spades) $= 4\cdot\dfrac{\binom{13}{5}}{\binom{52}{5}} = .00198$

$p(2) = P$ (all hearts and spades with at least one of each) + ...+ P (all diamonds and clubs with at least one of each)

$= 6P$ (all hearts and spades with at least one of each)

$= 6$ [P (1 heart $\cap$ 4 spades) + P (2 hearts $\cap$ 3 spades) + p (3 hearts $\cap$ 2 spades) + P (4 hearts $\cap$ 1 spade)]

$$= 6\left[2\cdot\frac{\binom{13}{4}\binom{13}{1}}{\binom{52}{5}} + 2\cdot\frac{\binom{13}{3}\binom{13}{2}}{\binom{52}{5}}\right] = 6\left[\frac{18{,}590 + 44{,}616}{2{,}598{,}960}\right] = 0.14592$$

$p(4) = 4P$ (2 spades, 1 heart, 1 diamond, 1 club)

$$= \frac{4\binom{13}{2}(13)(13)(13)}{\binom{52}{5}} = .26375$$

$p(3) = 1 - p(1) - p(2) - p(4) = .58835$

b $= \sum_{x=1}^{4} x\cdot p(x) = 3.12, \sigma^2 = \sum_{x=1}^{4} x^2 p(x) - (3.12)^2 = .384, \sigma = .620$

91 a $b(x;15,.75)$

b $B(12;15,.75) = .764$

c $B(12;15,.75) - B(8;15,.75) = .707$

d $\mu = (15)(.75) = 11.25, \sigma^2 = (15)(.75)(.25) = 2.81$

e Requests can all be met if and only if $X \le 10$ and $15 - X \le 8$, i.e. if $X \le 10$ and $7 \le X$, i.e. if $7 \le X \le 10$, so P (all requests met) $= B(10;15,.75) - B(6;15,.75) = .310$

93 Let $X \sim Bin\ (5, .9)$. Then $P(3 \le X) = 1 - P(X \le 2) = 1 - B(2;5,.9) = .991$

95

a $n = 500, p = .001$, so $np = .5$ and $b(x;500,.001) \doteq p(x;.5)$, a Poisson p.m.f.

b $P(X = 2) = p(2;.5) = F(2;.5) - F(1;.5) = .076$

c $P(2 \le X) = 1 - P(X \le 1) \doteq 1 - F(1;.5) = .090$

97 Let Y denote the number of tests carried out. For $n = 3$, possible Y values are 1 and 4. $P(Y = 1) = P$ (no one has the disease)

$= (.9)^3 = .729$ and $P(Y = 4) = .271$, so $E(Y) = (1)(.729) + (4)(.271) = 1.813$, as contrasted with the three tests necessary without group testing.

For $n = 5, E(Y) = (1)(.9)^5 + (6)[1 - (.9)^5] = 4.624$.

99 $p(2) = P(X = 2) = P$ (S on #1 and S on #2) $= p^2$

$p(3) = P$ (S on #3 and S on #2 and F on #1) $E = (1-p)p^2$

$p(4) = P$ (S on #4 and S on #3 and F on #2) $E = (1-p)p^2$

$p(5) = P$ (S on #5 and S on #4 and F on #3 and no two consecutive S's on trials prior to #3) $= [1 - p(2)](1-p)p^2$

$p(6) = P$ (S on #6 and S on #5 and F on #4 and no two consecutive S's on trials prior to #4) $= [1 - p(2) - p(3)](1-p)p^2$

In general, for $x = 5,6,7,\ldots,$

$p(x) = [1 - p(2) - \ldots - p(x-3)](1-p)p^2$

For $p = .9, p(2) = .81, p(3) = .081, p(4) = .081, p(5) = .01539,$

$p(6) = .008829, p(7) = .002268, p(8) = .00102141,$

so $P(X \le 8) = p(2) + \ldots + p(8) = .99950841$.

101

a $E(X) = np = (20)(.8) = 16$

b $20 - X =$ the number under warranty, so $E(20 - X) = 20 - E(X) = 4$

c At least 75% of 20 is at least 15, so $P(15 \le X) = 1 - B(14;20,.8) = .804$

d The hypergeometric pmf with population size $N = 12$, the number of population successes $M = 4$ (those under warranty are successes here), and sample size $n = 5$. Thus $E(X) = 1.67$ and $V(X) = \left[\dfrac{(7)(5)(4)(8)}{(11)(12)(12)}\right] = .707$

Chapter 3

103 **a** $P(X = 0) = F(0;2) = 0.135$

b Let $S =$ an operator who receives no requests. Then $p = .135$ and we wish

P (4 S's in 5 trials) $= b(4;5,.135) = \binom{5}{4}(.135)^4(.865)^1 = .00144$

c P (all receive x) = P (first receives x) $\cdot \ldots \cdot$ P (fifth receives x) $= [\frac{e^{-2}2^x}{x!}]^5$,

and P (all receive the same number) is the sum from $x = 0$ to ∞.

105 The number sold is min (X, 5), so $E[\min(X,5)] = \sum^{\infty} \min(x,5)\,p(x;4)$

$$= (0)p(0;4) + (1)p(1;4) + (2)p(2;4) + (3)p(3;4) + 4p(4;4)$$

$$+5 \sum_{x=5}^{\infty} p(x;4) = 1.735 + 5[1 - F(4;4)] = 3.59$$

107 **a** No; probability of success is not the same for all tests.

b There are four ways exactly three could have positive results. Let D represent those with the disease and D' represent those without the disease.

Combination		**Probability**
D'	D	
0	3	$\left[\binom{5}{0}(.2)^0(.8)^5\right]\cdot\left[\binom{5}{3}(.9)^3(.1)^2\right]$ $(.32768)(.0729) = .02389$
1	2	$\left[\binom{5}{1}(.2)^1(.8)^4\right]\cdot\left[\binom{5}{2}(.9)^2(.1)^3\right]$ $=(.4096)(.0081) = .00332$
2	1	$\left[\binom{5}{2}(.2)^2(.8)^3\right]\cdot\left[\binom{5}{1}(.9)^1(.1)^4\right]$ $(.2048)(.00045) = .00009216$
3	0	$\left[\binom{5}{3}(.2)^3(.8)^2\right]\cdot\left[\binom{5}{0}(.9)^0(.1)^5\right]$ $= (.0512)(.00001) = .000000512$

Adding up the probabilities associated with the four combinations yields 0.0273.

Chapter 3

109 **a** $p(x;\lambda,\mu) = \frac{1}{2}p(x;\lambda) + \frac{1}{2}p(x;\mu)$ where both $p(x;\mu)$ and $p(x;\lambda)$ are Poisson pmf's, and thus ≥ 0, so $p(x;\lambda,\mu) \geq 0$. Further,

$$\sum_{x=0}^{\infty} p(x;\lambda,\mu) = \frac{1}{2}\sum_{x=0}^{\infty} p(x;\lambda) + \frac{1}{2}\sum_{x=0}^{\infty} p(x;\mu) = \frac{1}{2} + \frac{1}{2} = 1$$

b $.6p(x;\lambda) + .4p(x;\mu)$

c $$E(X) = \sum_{x=0}^{\infty} x\{\frac{1}{2}p(x;\lambda) + \frac{1}{2}p(x;\mu)\} = \frac{1}{2}\sum_{x=0}^{\infty} xp(x;\lambda) + \frac{1}{2}\sum_{x=0}^{\infty} xp(x;\mu)$$
$$= \frac{1}{2}\lambda + \frac{1}{2}\mu = \frac{\lambda+\mu}{2}$$

d $E(X^2) = \frac{1}{2}\sum_{x=0}^{\infty} x^2p(x;\lambda) + \frac{1}{2}\sum_{x=0}^{\infty} x^2p(x;\mu) = \frac{1}{2}(\lambda^2+\lambda) + \frac{1}{2}(\mu^2+\mu)$ (since for a Poisson r.v., $E(X^2) = V(X) + [E(X)]^2 = \lambda + \lambda^2$),

so $V(X) = \frac{1}{2}[\lambda^2+\lambda+\mu^2+\mu] - [\frac{\lambda+\mu}{2}]^2 = \left(\frac{\lambda-\mu}{2}\right)^2 + \frac{\lambda+\mu}{2}$

111 $$P(X = j) = \sum_{i=1}^{10} P(\text{arm on track } i \cap X = j) = \sum_{i=1}^{10} P(X = j \mid \text{arm on } i) \cdot p_i$$
$$= \sum_{i=1}^{10} P(\text{next seek at } i+j+1 \text{ or } i-j-1) \cdot p_i = \sum_{i=1}^{10} (p_{i+j+1} + p_{i-j-1})p_i$$
where $p_k = 0$ if $k < 0$ or $k > 10$

113 Let $A = \{x: |x-\mu| \geq k\sigma\}$. Then

$\sigma^2 = \sum (x-\mu)^2p(x) \geq \sum_A (x-\mu)^2p(x) \geq (k\sigma)^2 \sum_A p(x)$

But $\sum_A p(x) = P(X \text{ is in } A) = P(|X-\mu| \geq k\sigma)$,

so $\sigma^2 \geq k^2\sigma^2 \cdot P(|X-\mu| \geq k\sigma)$ as desired.

CHAPTER 4

Section 4.1

1 a $P(X \le 1) = \int_{-\infty}^{1} f(x)dx = \int_0^1 \frac{1}{2}x\,dx = \frac{1}{4}x^2\Big]_0^1 = .25$

b $P(.5 \le X \le 1.5) = \int_{.5}^{1.5} \frac{1}{2}x\,dx = \frac{x^2}{4}\Big]_{.5}^{1.5} = .5$

c $P(1.5 < X) = \int_{1.5}^{\infty} f(x)dx = \int_{1.5}^{2} \frac{1}{2}x\,dx = \frac{x^2}{4}\Big]_{1.5}^{2} = \frac{7}{16} \approx .438$

3 a

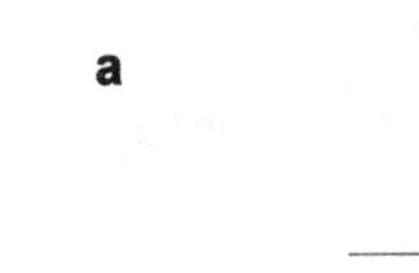

b $P(X > 0) = \int_0^1 \frac{3}{4}(1-x^2)dx = \frac{3}{4}\left(x - \frac{x^3}{3}\right)\Big]_0^1 = .5$

c $P(-.5 < X < .5) = \int_{-.5}^{.5} \frac{3}{4}(1-x^2)dx = \frac{3}{4}\left(x - \frac{x^3}{3}\right)\Big]_{-.5}^{.5} = \frac{33}{48} \approx .6875$

d $P(X < -.25 \text{ or } X > .25) = 1 - P(-.25 \le X \le .25) = 1 - \int_{-.25}^{.25} \frac{3}{4}(1-x^2)dx$

$$= 1 - \frac{3}{4}\left(x - \frac{x^3}{3}\right)\Big]_{-.25}^{.25} = \frac{81}{128} \approx .6328$$

5 a $1 = \int_{-\infty}^{\infty} f(x)dx = \int_0^1 kx^2dx = \frac{kx^3}{3}\Big]_0^1 = \frac{k}{3} \Rightarrow k = 3$

b $P\left(0 \le X \le \frac{1}{2}\right) = \int_0^{1/2} 3x^2dx = x^3\Big]_0^{1/2} = \frac{1}{8} = .125$

c $P\left(\frac{1}{4} \le X \le \frac{1}{2}\right) = \int_{1/4}^{1/2} 3x^2dx = x^3\Big]_{1/4}^{1/2} = \frac{7}{64} \approx .109$

d $P\left(\frac{2}{3} \le X\right) = \int_{2/3}^{1} 3x^2dx = x^3\Big]_{2/3}^{1} = \frac{19}{27} \approx .704$

7 a $f(x) = .1$ for $25 \le x \le 35$ and $= 0$ otherwise

b $P(X > 33) = \int_{33}^{35} \frac{1}{10}dx = .2$

Chapter 4

c $E(X) = \int_{25}^{35} x \cdot \frac{1}{10}\, dx = \frac{x^2}{20}\Big]_{25}^{35} = 30$

30 ± 2 is from 28 to 32 minutes

$P(28 < X < 32) = \int_{28}^{32} \frac{1}{10}\, dx = \frac{1}{10}x\Big]_{28}^{32} = 0.4$

d $P(a \le X \le a+2) = \int_{a}^{a+2} \frac{1}{10} dx = .2$ since the interval has length 2.

9 **a** $P(X \le 6) = \int_{.5}^{6} .15e^{-.15(x-.5)} dx = .15\int_{0}^{5.5} e^{-.15u} du$ (after $u = x - .5$)

$= -e^{-.15u}\Big]_{0}^{5.5} = 1 - e^{-.825} \approx .562$

b $1 - .562 = .438$; .438

c $P(5 \le X \le 6) = P(X \le 6) - P(X \le 5) \approx .562 - .491 = .071$

Section 4.2

11 **a** $P(X \le 1) = F(1) = \frac{1}{4} = .25$

b $P(.5 \le X \le 1) = F(1) - F(.5) = \frac{3}{16} = .1875$

c $P(X > .5) = 1 - P(X \le .5) = 1 - F(.5) = \frac{15}{16} = .9375$

d $.5 = F(\tilde{\mu}) = \frac{\tilde{\mu}^2}{4} \Rightarrow \tilde{\mu}^2 = 2 \Rightarrow \tilde{\mu} = \sqrt{2} \approx 1.4142$

e $f(x) = F'(x) = \frac{x}{2}$ for $0 \le x < 2$ and $= 0$ otherwise

13 **a** $E(X) = \int_{-\infty}^{\infty} x \cdot f(x)\, dx = \int_{0}^{2} x \cdot \frac{1}{2}x\, dx = \frac{1}{2}\int_{0}^{2} x^2\, dx = \frac{x^3}{6}\Big]_{0}^{2} = \frac{8}{6} \approx 1.33$

b $E(X^2) = \int_{-\infty}^{\infty} x^2 \cdot f(x)\, dx = \int_{0}^{2} x^2 \cdot \frac{1}{2}x\, dx = \frac{1}{2}\int_{0}^{2} x^3\, dx = \frac{x^4}{8}\Big]_{0}^{2} = 2,$

so $\text{Var}(X) = E(X^2) - [E(X)]^2 = 2 - \left(\frac{8}{6}\right)^2 = \frac{8}{36} \approx .222$, $\sigma_X \approx .471$

c From **b**, $E(X^2) = 2$.

Chapter 4

15 **a** $F(x) = 0$ for $x < 0$ and $= 1$ for $x > 1$. For $0 \le x \le 1$,

$$F(x) = \int_{-\infty}^{x} f(y)\,dy = \int_0^x 2(1-y)\,dy = 2\left[y - \frac{y^2}{2}\right]_0^x = 2\left[x - \frac{x^2}{2}\right].$$ Thus

$$F(x) = \begin{cases} 0 & x < 0 \\ 2\left[x - \dfrac{x^2}{2}\right] & 0 \le x \le 1 \\ 1 & 1 < x \end{cases}$$

1

1

b $$F(.5) = 2\left[\frac{1}{2} - \frac{\left(\frac{1}{2}\right)^2}{2}\right] = \frac{3}{4} = .75$$

c $$P(.25 < X \le .5) = F(.5) - F(.25) = \frac{3}{4} - \frac{7}{16} = \frac{5}{16} = .3125$$

$P(.25 \le X \le .5) = .3125$ also, since X is a continuous rv

d We must solve $F(x) = .75$ for x; $2\left[x - \frac{x^2}{2}\right] = .75 \Rightarrow x - \frac{x^2}{2} = \frac{3}{8}$

$\Rightarrow x^2 - 2x + \frac{3}{4} = 0 \Rightarrow x = \frac{1}{2}, \frac{3}{2}$ so $x = \frac{1}{2}$, or .5, is the 75th percentile.

e $$.5 = F(\tilde{\mu}) = 2\left[\tilde{\mu} - \frac{\tilde{\mu}^2}{2}\right] \Rightarrow \tilde{\mu}^2 - 2\tilde{\mu} + .5 = 0 \Rightarrow \tilde{\mu} = \frac{[2 - \sqrt{2}]}{2} \approx .293$$

f $$E(X) = \int_0^1 x \bullet 2(1-x)\,dx = 2\left[\frac{x^2}{2} - \frac{x^3}{3}\right]_0^1 = \frac{1}{3} \approx .333,$$

$$E(X^2) = \int_0^1 x^2 \bullet 2(1-x)\,dx = 2\left[\frac{x^3}{3} - \frac{x^4}{4}\right]_0^1 = \frac{1}{6} \approx .167, \text{ so}$$

$$\text{Var}(X) = \frac{1}{6} - \frac{1}{9} = \frac{1}{18} \approx .0556, \quad \sigma_X \approx .2357$$

Chapter 4

17 **a** For $2 \le x \le 4$, $F(x) = \int_{-\infty}^{x} f(y)\,dy = \int_{2}^{x} \frac{3}{4}[1-(y-3)^2]dy$ (let $u = y-3$)

$$= \int_{-1}^{x-3} \frac{3}{4}(1-u^2)du = \frac{3}{4}\left[u - \frac{u^3}{3}\right]_{-1}^{x-3} = \frac{3}{4}\left[x - \frac{7}{3} - \frac{(x-3)^3}{3}\right].$$ Thus

$$F(x) = \begin{cases} 0 & x < 2 \\ \frac{1}{4}[3x-7-(x-3)^3] & 2 \le x \le 4 \\ 1 & 4 < x \end{cases}$$

b By symmetry of $f(x)$, $\tilde{\mu} = 3$.

c $E(X) = \int_{2}^{4} x \cdot \frac{3}{4}[1-(x-3)^2]dx = \frac{3}{4}\int_{-1}^{1}(y+3)(1-y^2)dy$

$$= \frac{3}{4}\left[3y + \frac{y^2}{2} - y^3 - \frac{y^4}{4}\right]_{-1}^{1} = \frac{3}{4} \cdot 4 = 3$$

$\mathrm{Var}(X) = \int_{-\infty}^{\infty}(x-\mu)^2 f(x)\,dx = \frac{3}{4}\int_{2}^{4}(x-3)^2 \cdot [1-(x-3)^2]dx$

$$= \frac{3}{4}\int_{-1}^{1} y^2(1-y^2)\,dy = \frac{3}{4} \cdot \frac{4}{15} = \frac{1}{5} = .2$$

19 **a** $P(X \le 1) = F(1) = .25[1 + \ln(4)] \approx .597$

b $P(1 \le X \le 3) = F(3) - F(1) \approx .966 - .597 \approx .369$

c $f(x) = F'(x) = .25\ln(4) - .25\ln(x)$ for $0 < x < 4$

21 $E(\text{area}) = E(\pi R^2) = \int_{-\infty}^{\infty} \pi r^2 f(r)\,dr = \int_{9}^{11} \pi r^2 \left(\frac{3}{4}\right)[1-(10-r)^2]dr$

$$= \frac{3}{4}\pi\int_{9}^{11} r^2[1-(100-20r+r^2)]dr = \frac{3}{4}\pi\int_{9}^{11} -99r^2 + 20r^3 - r^4\,dr = 100.2\pi$$

23 With X = temperature in °C, temperature in °F = $\frac{9}{5}X + 32$,

so $E\left[\frac{9}{5}X+32\right] = \frac{9}{5}(120) + 32 = 248$, $\mathrm{Var}\left[\frac{9}{5}X+32\right] = \left(\frac{9}{5}\right)^2 \cdot (2)^2 = 12.96$,
so st. dev. = 3.6.

25 **a** $P(Y \le 1.8\tilde{\mu}+32) = P(1.8X+32 \le 1.8\tilde{\mu}+32) = P(X \le \tilde{\mu}) = .5$

b 90th for $Y = 1.8\eta(.9) + 32$ where $\eta(.9)$ is the 90th percentile for X, since

$$P(Y \le 1.8\eta(.9)+32) = P(1.8X+32 \le 1.8\eta(.9)+32) = P(X \le \eta(.9)) = .9$$ as desired.

c The (100p)th percentile for Y is $1.8\eta(p)+32$, verified by substituting p for .9 in the argument of **b**. When $Y = aX+b$ (i.e., a linear transformation of X), and the (100p)th percentile of the X distribution is $\eta(p)$, then the corresponding (100p)th percentile of the Y distribution is $a\bullet\eta(p)+b$ (i.e., same linear transformation applied to X's percentile).

Section 4.3

27 **a** .9838 is found in the 2.1 row and the .04 column of the standard normal table, so $c = 2.14$.

b $P(0 \le Z \le c) = .291 \Rightarrow \Phi(c) = .7910 \Rightarrow c = .81$

c $P(c \le Z) = .121 \Rightarrow 1-P(c \le Z) = P(Z < c) = \Phi(c) = 1-.121 = .8790$
$\Rightarrow c = 1.17$

d $P(-c \le Z \le c) = \Phi(c)-\Phi(-c) = \Phi(c)-[1-\Phi(c)] = 2\Phi(c)-1 = .668$
$\Rightarrow \Phi(c) = .8340 \Rightarrow c = .97$

e $P(c \le |Z|) = .016 \Rightarrow 1-.016 = .9840 = 1-P(c \le |Z|) = P(|Z| < c)$
$= P(-c < Z < c) = \Phi(c)-\Phi(-c) = 2\Phi(c)-1$
$\Rightarrow \Phi(c) = .9920 \Rightarrow c = 2.41$

29 **a** Area under Z curve above $z_{.0055}$ is .0055, which implies that
$\Phi(z_{.0055}) = 1-.0055 = .9945$, so $z_{.0055} = 2.54$

b $\Phi(z_{.09}) = .9100 \Rightarrow z = 1.34$ (since .9099 appears as the 1.34 entry).

c $\Phi(z_{.633})$ = area below $z_{.663} = .3370 \Rightarrow z_{.663} \approx -.42$

31 **a** $P(X \le 17) = P\left(Z \le \frac{17-15}{1.25}\right) = P(Z \le 1.60) = \Phi(1.60) = .9452$

b $P(12 \le X \le 17) = P(-2.40 \le Z \le 1.60) = .9370$

c $P(|X-15| \le 2(1.25)) = P(-2.5 \le X-15 \le 2.5) = P(12.5 \le X \le 17.5)$
$= P(-2.00 \le Z \le 2.00) = .9544$

33 **a** $P(X \le 14.8) = P(Z \le -1) = .1587$

b $P(14.7 \le X \le 15.1) = P(-1.5 \le Z \le .5) = .6247$

c P(travel 370 miles without refueling) = P(tank holds at least $\frac{370}{25}$ gallons)
$= P(14.8 \le X) = P(-1 \le Z) = .8413$

35 **a** $\mu+\sigma\bullet$(91st for stnd. normal) $= 25+5\bullet(1.34) = 31.7$

b $25+5\bullet(-1.555) = 17.225$

c $\mu = 3.000\ \mu m;\quad \sigma = 0.150$
We desire the 90th percentile
$3.0 + (1.28)(0.15) = 3.192$

37 $P(\text{damage}) = P(X < 100) = P\left(Z < \dfrac{100-200}{30}\right) = P(Z < -3.33) = .0004$
$P(\text{at least one among five is damaged}) = 1 - P(\text{none are damaged})$
$= 1-(.9996)^5 = 1-.998 = .002$

39 Since 1.28 is the 90th z percentile ($z_{.1} = 1.28$) and -1.645 is the 5th z percentile ($z_{.05} = 1.645$), the given information implies that $\mu + \sigma(1.28) = 10.256$ and $\mu + \sigma(-1.645) = 9.671$, from which $\sigma(-2.925) = -.585$, $\sigma = .2000$, and $\mu = 10$.

41
a $P(30.5 < X) = P\left(\dfrac{30.5-31}{.2} < Z\right) = P(-2.5 < Z) = .9938$
b $P(30.5 < X < 31.5) = P(-2.5 < Z < 2.5) = .9876$
c $P(X < 30.4) = P(Z < -3) = .0013$,
so $P(\text{at least one}) = 1 - P(\text{none}) = 1-(.9987)^4 = .0052$.

43 The stated condition implies that 99% of the area under the normal curve with $\mu = 10$ and $\sigma = 2$ is to the left of $c-1$, so $c-1$ is the 99th percentile of the distribution. Thus $c-1 = \mu + \sigma(2.33) = 14.66$, and $c = 15.66$.

45 $P(|X-\mu| \geq \sigma) = P(X \leq \mu - \sigma \text{ or } X \geq \mu + \sigma)$
$= 1 - P(\mu - \sigma \leq X \leq \mu + \sigma) = 1 - P(-1 \leq Z \leq 1) = .3174$
Similarly, $P(|X-\mu| \geq 2\sigma) = 1 - P(-2 \leq Z \leq 2) = .0456$,
and $P(|X-\mu| \geq 3\sigma) = 1 - P(-3 \leq Z \leq 3) = .0026$.

47

p:	.5	.6	.8
μ:	12.5	15	20
σ:	2.50	2.45	2.00

a

$P(15 \leq X \leq 20)$ =	$P(14.5 \leq \text{normal} \leq 20.5)$
.5: .212	$P(.80 \leq Z \leq 3.20) = .2112$
.6: .577	$P(-.20 \leq Z \leq 2.24) = .5668$
.8: .573	$P(-2.75 \leq Z \leq .25) = .5957$

b

$P(X \leq 15)$ =	$P(\text{normal} \leq 15.5)$
.885	$P(Z \leq 1.20) = .8849$
.575	$P(Z \leq .20) = .5793$
.017	$P(Z \leq -2.25) = .0122$

c

$P(20 \le X)$	=	$P(19.5 \le \text{normal})$
.002		.0026
.029		.0329
.617		.5987

49 $n = 500,\ p = .4,\ \mu = 200,\ \sigma = 10.9545$

a $P(180 \le X \le 230) = P(179.5 \le \text{normal} \le 230.5)$
$= P(-1.87 \le Z \le 2.78) = .9666$

b $P(X < 175) = P(X \le 174) = P(\text{normal} \le 174.5) = P(Z \le -2.33) = .0099$

51 **a** $F_Y(y) = P(Y \le y) = P(aX + b \le y) = P\left(X \le \dfrac{(y-b)}{a}\right)$ (for $a > 0$)

$$= \int_{-\infty}^{(y-b)/a} \frac{1}{\sqrt{2\pi}\,\sigma} e^{-(1/2\sigma^2)(x-\mu)^2} dx$$

Now differentiate with respect to y to obtain

$f_Y(y) = F_Y'(y) = \dfrac{1}{\sqrt{2\pi}\,a\sigma} e^{-\frac{1}{2a^2\sigma^2}[y-(a\mu+b)]^2}$, so Y is normal with mean $a\mu + b$ and variance $a^2\sigma^2$.

b Normal, mean $\dfrac{9}{5}(115) + 32 = 239$, variance = 12.96

Section 4.4

53 **a** $\Gamma(6) = 5! = 120$

b $\Gamma\left(\dfrac{5}{2}\right) = \dfrac{3}{2}\Gamma\left(\dfrac{3}{2}\right) = \dfrac{3}{2}\bullet\dfrac{1}{2}\bullet\Gamma\left(\dfrac{1}{2}\right) = \left(\dfrac{3}{4}\right)\sqrt{\pi} \approx 1.329$

c $F(4;5) = .371$ from row 4, column 5 of Table A.4

d $F(5;4) = .735$

e $F(0;4) = P(X \le 0;\ \alpha = 4) = 0$

55 **a** $P(X \le 1) = F\left(\dfrac{1}{1/2};\ 2\right) = F(2;2) = .594$

b $P(2 < X) = 1 - P(X \le 2) = 1 - F\left(\dfrac{2}{1/2};2\right) = 1 - F(4;\ 2) = .092$

c $P(.5 \le X \le 1.5) = F(3;\ 2) - F(1;\ 2) = .537$

57 $\mu = 24,\ \sigma^2 = 144 \Rightarrow \alpha\beta = 24,\ \alpha\beta^2 = 144 \Rightarrow \beta = 6,\ \alpha = 4$

a $P(12 \le X \le 24) = F(4;\ 4) - F(2;\ 4) = .424$

b $P(X \le 24) = F(4;\, 4) = .567$, so while the mean is 24, the median is less than $24 (P(X \le \tilde{\mu}) = .5)$; this is a result of the positive skew of the gamma distribution.

c We want c such that $.99 = P(X \le c) = F\left(\frac{c}{6};\, 4\right)$. From the $\alpha = 4$ column of Table A.4, .990 appears in row 10, so $\frac{c}{6} = 10$, which implies that $c = 60$.

d The desired value of t is the 99.5th percentile, so

$$.995 = F\left(\frac{t}{6};\, 4\right) \Rightarrow \frac{t}{6} = 11 \Rightarrow t = 66.$$

59 a $P(X \le 30) = 1 - e^{-\lambda \cdot (30)} = 1 - e^{-(1/20) \cdot 30} = 1 - e^{-1.5} \approx .777$

$\left(\text{since } E(X) = \frac{1}{\lambda} = 20\right)$

b $P(X \ge 20) = 1 - \left[1 - e^{-(1/20) \cdot 20}\right] = e^{-1} \approx .368$

c $P(20 \le X \le 30) \approx .777 - .632 = .145$

d $1 - e^{-t/20} = .5 \Rightarrow e^{-t/20} = .5 \Rightarrow \frac{-t}{20} = \ln(.5) = -.693 \Rightarrow t = 13.863$

61 a $E(X) = \alpha\beta = n\frac{1}{\lambda} = \frac{n}{\lambda}$; for $\lambda = .5$, $n = 10$, $E(X) = 20$

b $P(X \le 30) = F\left(\frac{30}{2};10\right) = F(15;10) = .930$

c $P(X \le t) = P(\text{at least } n \text{ events in time } t) = P(Y \ge n)$ when $Y \sim$ Poisson with parameter λt. Thus $P(X \le t) = 1 - P(Y < n) = 1 - P(Y \le n-1)$

$$= 1 - \sum_{k=0}^{n-1} \frac{e^{-\lambda t}(\lambda t)^k}{k!}.$$

63 With x_p = (100p)th percentile, $p = F(x_p) = 1 - e^{-\lambda x_p} \Rightarrow e^{-\lambda x_p} = 1 - p$

$\Rightarrow -\lambda x_p = \ln(1-p) \Rightarrow x_p = \frac{-[\ln(1-p)]}{\lambda}$. For $p = .5$, $x_{.5} = \tilde{\mu} = \frac{.693}{\lambda}$.

65 a $\{X^2 \le y\} = \{-\sqrt{y} \le X \le \sqrt{y}\}$

b $P(X^2 \le y) = \int_{-\sqrt{y}}^{\sqrt{y}} \frac{1}{\sqrt{2\pi}} e^{-z^2/2}\, dz$. Now differentiate with respect to y to obtain the chi-squared pdf with $\nu = 1$.

Section 4.5

67 a $P(X \le 200) = F(200;\,2.5,\,200) = 1 - e^{-(200/200)^{2.5}} = 1 - e^{-1} \approx .632$

$P(X < 200) = P(X \le 200) \approx .632$

$P(X > 300) = 1 - F(300;\,2.5,\,200) = e^{-(1.5)^{2.5}}$

b $P(100 \le X \le 200) = F(200;\,2.5,\,200) - F(100;\,2.5,\,200)$

$\approx .632 - .162 = .470$

c The median $\tilde{\mu}$ is requested. The equation $F(\tilde{\mu}) = .5$ reduces to

$.5 = e^{-(\tilde{\mu}/200)^{2.5}}$, i.e., $\ln(.5) \approx -\left(\dfrac{\tilde{\mu}}{200}\right)^{2.5}$, so $\tilde{\mu} = (.6931)^{.4}(200) = 172.727$.

69 $\mu = \int_0^\infty x \bullet \dfrac{\alpha}{\beta^\alpha} x^{\alpha-1} e^{-(x/\beta)^\alpha}\,dx = \left(\text{after } y = \left(\dfrac{x}{\beta}\right)^\alpha,\ dy = \dfrac{\alpha x^{\alpha-1}}{\beta^\alpha}dx\right)$

$\beta \int_0^\infty y^{1/\alpha} e^{-y}\,dy = \beta \bullet \Gamma\left(1 + \dfrac{1}{\alpha}\right)$ by definition of the gamma function.

71 a $P(X \le 100) = P(\ln(X) \le \ln(100)) = \Phi\left(\dfrac{\ln(100) - 3.5}{1.2}\right) \approx \Phi(.92) = .8212$

b $P(100 \le X \le 200)) = P(\ln(100) \le \ln(X) \le \ln(200))$

$= \Phi\left(\dfrac{\ln(200) - 3.5}{1.2}\right) - \Phi\left(\dfrac{\ln(100) - 3.5}{1.2}\right) \approx \Phi(1.50) - \Phi(.92) = .1120$

73 a $E(X) = e^{5+(.01)/2} = e^{5.005} = 149.157$, $\text{Var}(X) = e^{10+.01} \bullet (e^{.01} - 1) = 223.594$

b $P(X > 120) = 1 - P(X \le 120) = 1 - P(\ln(X) \le 4.788) = 1 - \Phi(-2.12) = .9830$

c $P(110 \le X \le 130) = P(4.7005 \le \ln(X) \le 4.8675) = \Phi(-1.32) - \Phi(-3) = .0921$

d $\tilde{\mu} = e^5 = 148.41$.

e P(any particular one has $X > 120$) = .983 $\Rightarrow$ expected # = 10(.983) = 9.83

f We wish the 5th percentile, which is $e^{5+(-1.645)(.1)} = 125.90$.

75 The point of symmetry must be $\dfrac{1}{2}$, so we require that $f\left(\dfrac{1}{2} - u\right) = f\left(\dfrac{1}{2} + u\right)$, i.e.,

$\left(\dfrac{1}{2} - u\right)^{\alpha-1}\left(\dfrac{1}{2} + u\right)^{\beta-1} = \left(\dfrac{1}{2} + u\right)^{\alpha-1}\left(\dfrac{1}{2} - u\right)^{\beta-1}$, which in turn implies that $\alpha = \beta$.

77 **a**

$$E(X) = \int_0^1 x \bullet \frac{\Gamma(\alpha+\beta)}{\Gamma(\alpha)\Gamma(\beta)} x^{\alpha-1}(1-x)^{\beta-1}\,dx = \frac{\Gamma(\alpha+\beta)}{\Gamma(\alpha)\Gamma(\beta)}\int_0^1 x^{\alpha+1-1}(1-x)^{\beta-1}dx$$

$$= \frac{\Gamma(\alpha+\beta)}{\Gamma(\alpha)\Gamma(\beta)} \bullet \frac{\Gamma(\alpha+1)\Gamma(\beta)}{\Gamma(\alpha+\beta+1)} = \frac{\alpha\Gamma(\alpha)}{\Gamma(\alpha)} \bullet \frac{\Gamma(\alpha+\beta)}{(\alpha+\beta)\Gamma(\alpha+\beta)} = \frac{\alpha}{\alpha+\beta}$$

b

$$E[(1-X)^m] = \int_0^1 (1-x)^m \frac{\Gamma(\alpha+\beta)}{\Gamma(\alpha)\Gamma(\beta)} x^{\alpha-1}(1-x)^{\beta-1}\,dx$$

$$= \frac{\Gamma(\alpha+\beta)}{\Gamma(\alpha)\Gamma(\beta)}\int_0^1 x^{\alpha-1}(1-x)^{m+\beta-1}\,dx = \frac{\Gamma(\alpha+\beta)\bullet\Gamma(m+\beta)}{\Gamma(\alpha+\beta+m)\Gamma(\beta)}$$

For $m = 1$, $E(1-X) = \dfrac{\beta}{\alpha+\beta}$.

Section 4.6

79 The z percentiles are as given in Example 29. The accompanying plot is quite straight except for the point corresponding to the largest observation (1.645, 422.6). This observation is clearly much larger than what would be expected in a normal random sample. Because of this outlier, it would be inadvisable to analyze the data using any inferential method that depended on assuming a normal population distribution.

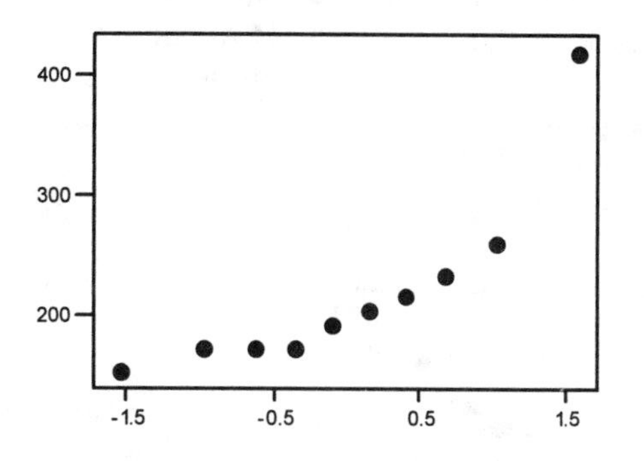

81 **a** The ten largest z percentiles are 1.96, 1.44, 1.15, .93, .76, .60, .45, .32, .19 and .06; the remaining ten are the negatives of these values. The accompanying normal probability plot is reasonably straight. An assumption of population distribution normality is plausible.

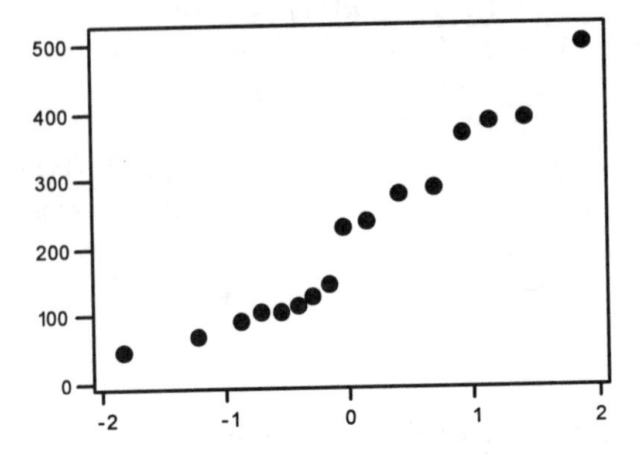

b For a Weibull probability plot, the natural logs of the observations are plotted against extreme value percentiles; these percentiles are −3.68, −2.55, −2.01, −1.65, −1.37, −1.13, −.93, −.76, −.59, −.44, −.30, −.16, −.02, .12, .26, .40, .56, .73, .95, and 1.31. The accompanying probability plot is roughly as straight as the one for checking normality (a plot of $\ln(x_i)$ versus the z percentiles, appropriate for checking the plausibility of a lognormal distribution, is also reasonably straight — any of three different families of population distributions seems plausible!).

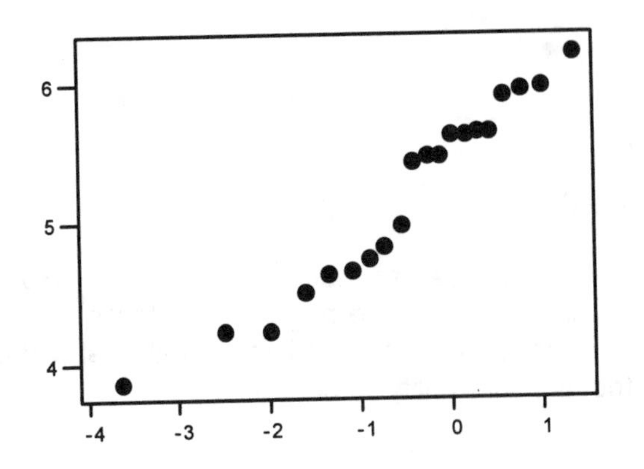

Chapter 4

83 To check for plausibility of a lognormal population distribution for the rainfall data of Exercise 67b in Chapter 1, take the natural logs and construct a normal probability plot. This plot and a normal probability plot for the original data appear below. Clearly the log transformation gives quite a straight plot, so lognormality is plausible. The curvature in the plot for the original data implies a positively skewed population distribution — like the lognormal distribution.

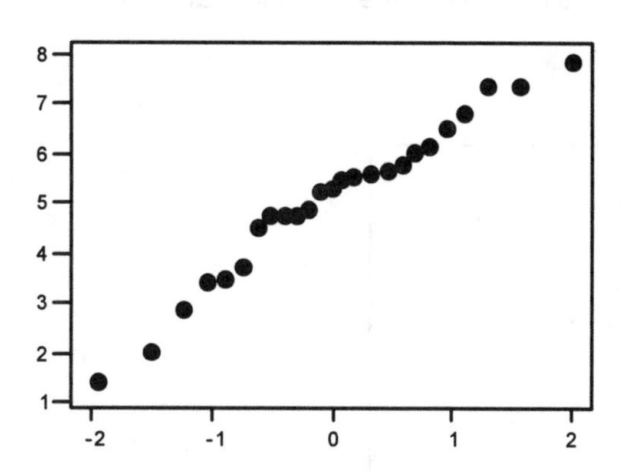

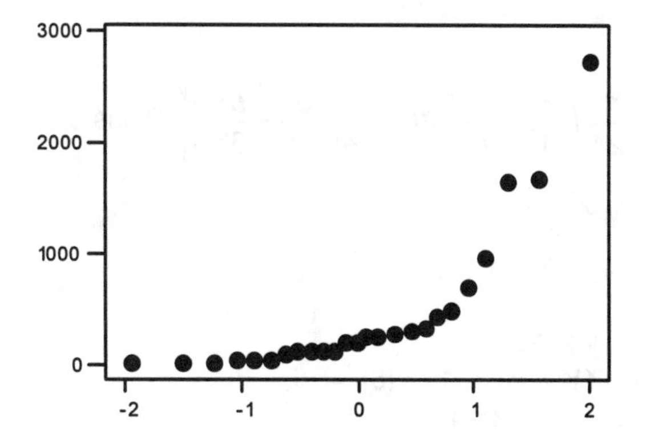

85 The (100p)th percentile $\eta(p)$ for the exponential distribution with $\lambda = 1$ satisfies $F(\eta(p)) = 1 - \exp[-\eta(p)] = p$, i.e., $\eta(p) = -\ln(1-p)$. With n = 16, we need $\eta(p)$ for $p = \frac{.5}{16}, \frac{1.5}{16}, \ldots, \frac{15.5}{16}$. These are .032, .098, .170, .247, .330, .421, .521, .633, .758, .901, 1.068, 1.269, 1.520, 1.856, 2.367, 3.466. This plot exhibits substantial curvature, casting doubt on the assumption of an exponential population distribution. Because λ is a scale parameter (as is σ for the normal family), $\lambda = 1$ can be used to assess the plausibility of the entire exponential family.

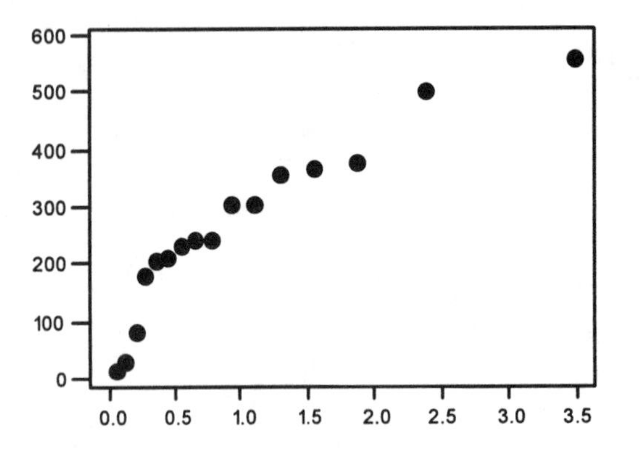

Supplementary

87 **a** For $0 \le y \le 12$, $F(y) = \frac{1}{24}\int_0^y \left(u - \frac{u^2}{12}\right)du = \frac{1}{24}\left(\frac{u^2}{2} - \frac{u^3}{36}\right)\Big]_0^y$. Thus

$$F(y) = \begin{cases} 0 & y < 0 \\ \frac{1}{48}\left(y^2 - \frac{y^3}{18}\right) & 0 \le y \le 12 \\ 1 & y > 12 \end{cases}$$

b $P(Y \le 4) = F(4) = .259$, $P(Y > 6) = 1 - F(6) = .5$, $P(4 \le Y \le 6) = F(6) - F(4) = .5 - .259 = .241$

c $E(Y) = \frac{1}{24}\int_0^{12} y^2\left(1-\frac{y}{12}\right)dy = \frac{1}{24}\left[\frac{y^3}{3}-\frac{y^4}{48}\right]_0^{12} = 6$

$E(Y^2) = \frac{1}{24}\int_0^{12} y^3\left(1-\frac{y}{12}\right)dy = 43.2$, so $V(Y) = 43.2-36 = 7.2$

d $P(Y < 4 \text{ or } Y > 8) = 1-P(4 \le Y \le 8) = .518$

e The shorter segment has length $\min(Y, 12-Y)$ so

$E[\min(Y, 12-Y)] = \int_0^{12}\min(y,12-y)\bullet f(y)dy = \int_0^{6}\min(y,12-y)\bullet f(y)\,dy$

$+\int_6^{12}\min(y,12-y)\bullet f(y)dy = \int_0^{6} y\bullet f(y)dy+\int_6^{12}(12-y)f(y)dy = \frac{90}{24} = 3.75.$

89 **a** By differentiation,

$$f(x) = \begin{cases} x^2 & 0 \le x \le 1 \\ \frac{7}{4}-\frac{3}{4}x & 1 \le x \le \frac{7}{3} \\ 0 & \text{otherwise} \end{cases}$$

b $P(.5 \le X \le 2) = F(2)-F(.5) = 1-\frac{1}{2}\left(\frac{7}{3}-2\right)\left(\frac{7}{4}-\frac{3}{4}\bullet 2\right)-\frac{(.5)^3}{3} = \frac{11}{12} = .917$

c $E(X) = \int_0^{1} x\bullet x^2dx+\int_1^{7/3} x\bullet\left(\frac{7}{4}-\frac{3}{4}x\right)dx = \frac{131}{108} = 1.213$

91 $\mu = 137.2$ oz., $\sigma = 1.6$ oz.

a $P(X > 135) = 1-\Phi\left(\frac{135-137.2}{1.6}\right) = 1-\Phi(-1.38)$

$= 1-0.0838 = 0.9162$

b With Y = the number among ten that contain more than 135 oz., $Y \sim \text{Bin}(10, .9162)$, so $P(Y \ge 8) = b(8; 10, .9162) + b(9; .9162) + b(10, .9162) = .9549$

c $\mu = 137.2$

$\frac{135-137.2}{\sigma} = -1.65$

$\sigma = 1.3374$

Chapter 4

93 **a**

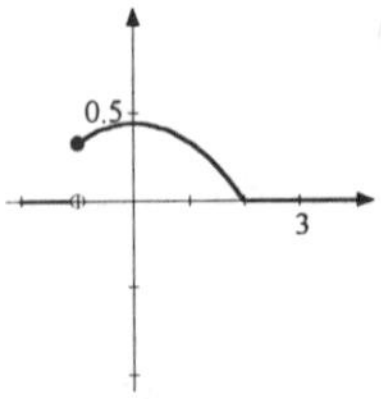

b $F(x) = 0$ for $x < -1$ or $= 1$ for $x > 2$. For $-1 \le x \le 2$,

$$F(x) = \int_{-1}^{x} \frac{1}{9}(4 - y^2)dy = \frac{1}{9}\left[4x - \frac{x^3}{3}\right] + \frac{11}{27}$$

c The median is 0 iff $F(0) = .5$. Since $F(0) = \frac{11}{27}$, this is not the case. Because $\frac{11}{27} < .5$, the median must be greater than 0.

d Y is a binomial rv with $n = 10$ and $p = P(X > 1) = 1 - F(1) = \frac{5}{27}$.

95 **a** $P(X \le 150) = \exp\left[-\exp\left(\frac{-(150-150)}{90}\right)\right] = \exp[-\exp(0)] = \exp(-1) = .368$,

where $\exp(u) = e^u$

$P(X \le 300) = \exp[-\exp(-1.6667)] = .828$, and

$P(150 \le X \le 300) = .828 - .368 = .460$

b The desired value c is the 90th percentile, so c satisfies

$.9 = \exp\left[-\exp\left(-\frac{(c-150)}{90}\right)\right]$. Taking the natural log of each side twice in succession yields $\ln[-\ln(.9)] = \frac{-(c-150)}{90}$,

so $c = 90(2.250367) + 150 = 352.53$.

c $f(x) = F'(x) = \frac{1}{\beta} \bullet \exp\left[-\exp\left(\frac{-(x-\alpha)}{\beta}\right)\right] \bullet \exp\left(\frac{-(x-\alpha)}{\beta}\right)$

d We wish the value of x for which $f(x)$ is a maximum; this is the same as the value of x for which $\ln[f(x)]$ is a maximum. The equation of $\frac{d[\ln(f(x))]}{dx} = 0$ gives $\exp\left(\frac{-(x-\alpha)}{\beta}\right) = 1$, so $\frac{-(x-\alpha)}{\beta} = 0$, which implies that $x = \alpha$. Thus the mode is α.

e $E(X) = .5772\beta + \alpha = 201.95$, whereas the mode is 150 and the median is $-(90)\ln[-\ln(.5)] + 150 = 182.99$. The distribution is positively skewed.

97

a From a graph of $f(x;\mu,\sigma)$ or by differentiation, $x^* = \mu$.

b No; the density function has constant height for $A \le x \le B$.

c $f(x;\lambda)$ is largest for $x = 0$ (the derivative at 0 does not exist since f is not continuous there) so $x* = 0$.

d $\ln[f(x;\alpha,\beta)] = -\ln(\beta^\alpha) - \ln(\Gamma(\alpha)) + (\alpha-1)\ln(x) - \dfrac{x}{\beta};$

$$\frac{d}{dx}\ln[f(x;\alpha,\beta)] = \frac{\alpha-1}{x} - \frac{1}{\beta} \Rightarrow x = x^* = (\alpha-1)\beta$$

e From **d**, $x^* = \left(\dfrac{\nu}{2}-1\right)(2) = \nu-2.$

99

a Clearly $f(x;\lambda_1,\lambda_2,p) \ge 0$ for all x, and

$$\int_{-\infty}^{\infty} f(x;\ \lambda_1,\lambda_2,p)dx = \int_0^{\infty}\left[p\lambda_1 e^{-\lambda_1 x} + (1-p)\lambda_2 e^{-\lambda_2 x}\right]dx$$

$$= p\int_0^{\infty}\lambda_1 e^{-\lambda_1 x}dx + (1-p)\int_0^{\infty}\lambda_2 e^{-\lambda_2 x}dx = p + (1-p) = 1$$

b For $x \ge 0$, $F(x;\ \lambda_1,\lambda_2,p) = \int_0^x f(y;\ \lambda_1,\lambda_2,p)dy = p(1-e^{-\lambda_1 x}) + (1-p)(1-e^{-\lambda_2 x})$.

c $E(X) = \int_0^{\infty} x\bullet[p\lambda_1 e^{-\lambda_1 x} + (1-p)\lambda_2 e^{-\lambda_2 x}]dx = p\int_0^{\infty} x\lambda_1 e^{-\lambda_1 x}dx + (1-p)\int_0^{\infty} x\lambda_2 e^{-\lambda_2 x}dx$

$$= \frac{p}{\lambda_1} + \frac{(1-p)}{\lambda_2}$$

d $E(X^2) = \dfrac{2p}{\lambda_1^2} + \dfrac{2(1-p)}{\lambda_2^2}$, so $\text{Var}(X) = \dfrac{2p}{\lambda_1^2} + \dfrac{2(1-p)}{\lambda_2^2} - \left[\dfrac{p}{\lambda_1} + \dfrac{(1-p)}{\lambda^2}\right]^2$

e For an exponential rv, $CV = \dfrac{1/\lambda}{1/\lambda} = 1$. For X hyperexponential,

$$CV = \left[\frac{\dfrac{2p}{\lambda_1^2} + \dfrac{2(1-p)}{\lambda_2^2}}{\left(\dfrac{p}{\lambda_1} + \dfrac{(1-p)}{\lambda_2}\right)^2} - 1\right]^{1/2} = \left[\frac{2(p\lambda_2^2 + (1-p)\lambda_1^2)}{(p\lambda_2 + (1-p)\lambda_1)^2} - 1\right]^{1/2}$$

$= [2r-1]^{1/2}$ where $r = \dfrac{p\lambda_2^2 + (1-p)\lambda_1^2}{[p\lambda_2 + (1-p)\lambda_1]^2}$. But straightforward algebra shows that $r > 1$ provided that $\lambda_1 \neq \lambda_2$, so that $CV > 1$.

f $\mu = \dfrac{n}{\lambda}$, $\sigma^2 = \dfrac{n}{\lambda^2}$, so $\sigma = \dfrac{\sqrt{n}}{\lambda}$ and $CV = \dfrac{1}{\sqrt{n}} < 1$ if $n > 1$.

101 **a** A lognormal distribution, since $\ln\left(\dfrac{I_o}{I_i}\right)$ is a normal rv

b $P(I_o > 2I_i) = P\left(\dfrac{I_o}{I_i} > 2\right) = P\left(\ln\left(\dfrac{I_o}{I_i}\right) > \ln 2\right) = 1 - P\left(\ln\left(\dfrac{I_o}{I_i}\right) \le \ln 2\right)$

$= 1 - \phi\left(\dfrac{\ln 2 - 1}{.05}\right) = 1 - \phi(-6.14) = 1$

c $E\left(\dfrac{I_o}{I_i}\right) = e^{1+.0025/2} = 2.72$, $\text{Var}\left(\dfrac{I_o}{I_i}\right) = e^{2+.0025} \bullet (e^{.0025} - 1) = .0185$

103 **a** $F_Y(y) = P(Y \le y) = P(60X \le y) = P\left(X \le \dfrac{y}{60}\right) = F\left(\dfrac{y}{60\beta}; \alpha\right)$. Thus

$f_Y(y) = f\left(\dfrac{y}{60\beta}; \alpha\right) \bullet \dfrac{1}{60\beta} = \dfrac{y^{\alpha-1} e^{\frac{-y}{60\beta}}}{(60\beta)^{\alpha}\Gamma(\alpha)}$, which shows that Y has a gamma distribution with parameters α and 60β.

b With c replacing 60 in **a**, the same argument shows that cX has a gamma distribution with parameters α and $c\beta$.

105 **a** $f(x) = \lambda e^{-\lambda x}$ and $F(x) = 1 - e^{-\lambda x}$, so $r(x) = \dfrac{\lambda e^{-\lambda x}}{e^{-\lambda x}} = \lambda$, a constant (independent of X); this is consistent with the memoryless property of the exponential distribution.

b $r(x) = \left(\dfrac{\alpha}{\beta^{\alpha}}\right) x^{\alpha-1}$; for $\alpha > 1$ this is increasing, while for $\alpha < 1$ it is a decreasing function.

c $\ln(1 - F(x)) = -\int \alpha\left(1 - \dfrac{x}{\beta}\right) dx = -\alpha\left[x - \dfrac{x^2}{2\beta}\right] \Rightarrow F(x) = 1 - e^{-\alpha\left(x - \frac{x^2}{2\beta}\right)}$,

$f(x) = \alpha\left(1 - \dfrac{x}{\beta}\right) e^{-\alpha\left(x - \frac{x^2}{2\beta}\right)} \quad 0 \le x \le \beta$

107 **a** $E(g(X)) \approx E[g(\mu) + g'(\mu)(X-\mu)] = E(g(\mu)) + g'(\mu) \bullet E(X-\mu)$, but $E(X) - \mu = 0$ and $E(g(\mu)) = g(\mu)$ (since $g(\mu)$ is constant), giving $E(g(X)) \approx g(\mu)$.
$V(g(X)) \approx V[g(\mu) + g'(\mu)(X-\mu)] = V[g'(\mu)(X-\mu)] = (g'(\mu))^2 \bullet V(X-\mu)$
$= [g'(\mu)]^2 \bullet V(X)$.

b $g(I) = \dfrac{v}{I}$, $g'(I) = \dfrac{-v}{I^2}$, so $E(g(I)) = \mu_R \approx \dfrac{v}{\mu_I} = \dfrac{v}{20}$

$$V(g(I)) \approx \left(\frac{-v}{\mu_I^2}\right)^2 \bullet V(I),\ \sigma_{g(I)} \approx \frac{v}{20^2} \bullet \sigma_I = \frac{v}{800}$$

109 For $y > 0$,

$$F(y) = P(Y \le y) = P\left(\frac{2X^2}{\beta^2} \le y\right) = P\left(X^2 \le \frac{\beta^2 y}{2}\right) = P\left(X \le \frac{\beta\sqrt{y}}{\sqrt{2}}\right).$$

Now take the cdf of X (Weibull), replace x by $\dfrac{\beta\sqrt{y}}{\sqrt{2}}$, and then differentiate with respect to y to obtain the desired result $f_Y(y)$.

CHAPTER 5

Section 5.1

1 **a** $P(X = 1, Y = 1) = p(1,1) = .20$

b $P(X \le 1 \text{ and } Y \le 1) = p(0,0) + p(0,1) + p(1,0) + p(1,1) = .42$

c At least one hose is in use at both islands. $P(X \ne 0 \text{ and } Y \ne 0)$
$= p(1,1) + p(1,2) + p(2,1) + p(2,2) = .70$

d By summing row probabilities, $p_X(x) = .16, .34, .50$ for $x = 0,1,2$, and by summing column probabilities, $p_Y(y) = .24, .38, .38$ for $y = 0,1,2$. $P(X \le 1) = p_X(0) + p_X(1) = .50$

e $p(0,0) = .10$, but $p_X(0) \cdot p_y(0) = (.16)(.24) = .0384 \ne .10$, so X and Y are not independent.

3 **a** $p(1,1) = .15$, the entry in the 1 row and 1 column of the joint probability table.

b $P(X_1 = X_2) = p(0,0) + p(1,1) + p(2,2) + p(3,3)$
$= .08 + .15 + .10 + .07 = .40$

c $A = \{(x_1, x_2) : x_1 \ge 2 + x_2\} \cup \{(x_1, x_2) : x_2 \ge 2 + x_1\}$
$P(A) = p(2,0) + p(3,0) + p(4,0) + p(3,1) + p(4,1) + p(4,2)$
$+ p(0,2) + p(0,3) + p(1,3)$
$= .22$

d $P(\text{exactly } 4) = p(1,3) + p(2,2) + p(3,1) + p(4,0) = 0.17$
$P(\text{at least } 4) = P(\text{exactly } 4) + p(4,1) + p(4,2) + p(4,3)$
$+ p(3,2) + p(3,3) + p(2,3)$
$= .46$

5 **a** $P(X = 3, Y = 3) = P$ (3 customers, each with 1 package)
$= P$ (each has 1 package | 3 customers)
• P (3 customers)
$= (.6)^3 \cdot (.25) = .054$

b $P(X = 4, Y = 11) = P$ (total of 11 packages | 4 customers)
• P (4 customers)
Given that there are 4 customers, there are 4 different ways to have a total of 11 packages: 3, 3, 3, 2 or 3, 3, 2, 3 or 3, 2, 3, 3 or 2, 3, 3, 3. Each way has probability $(.1)^3 \cdot (.3)$, so $p(4,11) = 4(.1)^3(.3)(.15) = .00018$

7 **a** $p(1,1) = .030$

b $P(X \le 1 \text{ and } Y \le 1) = p(0,0) + p(0,1) + p(1,0) + p(1,1) = .120$

c $P(X = 1) = p(1,0) + p(1,1) + p(1,2) = .100$;
$P(Y = 1) = p(0,1) + \ldots + p(5,1) = .300$

d $P(\text{overflow}) = P(X+3Y > 5) = 1 - P(X+3Y \le 5) = 1 - P((X,Y)$
$= (0,0)$ or .. or $(5,0)$ or $(0,1)$ or $(1,1)$ or $(2,1))$
$= 1 - .620 = .380$

e The marginal probabilities for X (row sums from the joint probability table) are $p_X(0) = .05, p_X(1) = .10, p_X(2) = .25,$
$p_X(3) = .30, p_X(4) = .20,$ and $p_X(5) = .10$; those for Y (column sums) are $p_Y(0) = .5, p_Y(1) = .3,$ and $p_Y(2) = .2$. It is now easily verified that for every $(x,y), p(x,y) = p_X(x) \cdot p_Y(y)$, so X and Y are independent.

9 a
$$1 = \int_{-\infty}^{\infty}\int_{-\infty}^{\infty} f(x,y)\,dx\,dy = \int_{20}^{30}\int_{20}^{30} K(x^2+y^2)\,dx\,dy$$
$$= K\int_{20}^{30}\int_{20}^{30} x^2\,dy\,dx + K\int_{20}^{30}\int_{20}^{30} y^2\,dx\,dy = 10K\int_{20}^{30} x^2 dx + 10K\int_{20}^{30} y^2 dy$$
$$= 20K \cdot \left(\frac{19{,}000}{3}\right) \Rightarrow K = \frac{3}{380{,}000}$$

b
$$P(X < 26 \text{ and } Y < 26) = \int_{20}^{26}\int_{20}^{26} K(x^2+y^2)\,dx\,dy = 12K\int_{20}^{26} x^2\,dx$$
$$= 4Kx^3\Big|_{20}^{26} = 38{,}304K = .3024$$

c

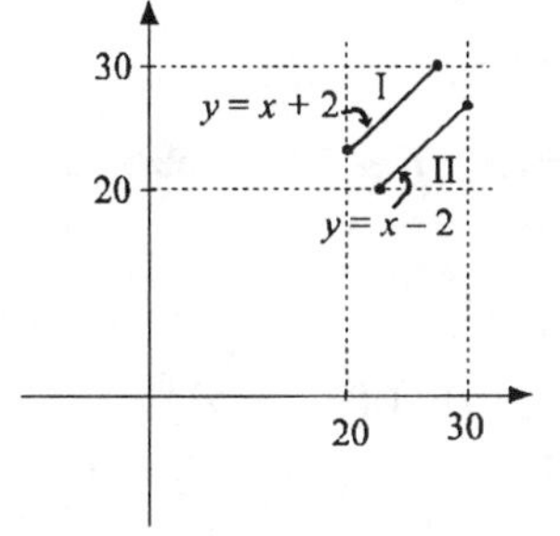

$$P(|X-Y| \le 2) = \iint_{\substack{\text{shaded}\\ \text{region}}} f(x,y)\,dx\,dy$$
$$= 1 - \iint f(x,y)\,dx\,dy - \iint f(x,y)\,dx\,dy$$
$$= 1 - \int_{20}^{28}\int_{x+2}^{30} f(x,y)\,dy\,dx - \int_{22}^{30}\int_{20}^{x-2} f(x,y)\,dy\,dx$$
$= $ (after much algebra) $.3593$

d
$$f_X(x) = \int_{-\infty}^{\infty} f(x,y)\,dy = \int_{20}^{30} K(x^2+y^2)\,dy = 10Kx^2 + K\frac{y^3}{3}\Big|_{20}^{30}$$
$$= 10Kx^2 + .05 \text{ for } 20 \le x \le 30$$

e $f_Y(y)$ is obtained by substituting y for x in (d); clearly $f(x,y) \ne f_X(x) \cdot f_Y(y)$, so X and Y are not independent.

Chapter 5

11 a $p(x,y) = \dfrac{e^{-\lambda}\lambda^x}{x!} \cdot \dfrac{e^{-\mu}\mu^y}{y!}$ for $x = 0,1,2,\ldots;\ y = 0,1,2,\ldots$

b $p(0,0) + p(0,1) + p(1,0) = e^{-\lambda-\mu}[1 + \lambda + \mu]$

c $P(X+Y = m) = \sum_{k=0}^{m} P(X = k,\ Y = m-k) = \sum_{k=0}^{m} e^{-\lambda-\mu}\dfrac{\lambda^k}{k!}\dfrac{\mu^{m-k}}{(m-k)!}$

$$= \frac{e^{-(\lambda+\mu)}}{m!}\sum_{k=0}^{m}\binom{m}{k}\lambda^k\mu^{m-k} = \frac{e^{-(\lambda+\mu)}(\lambda+\mu)^m}{m!},$$

so the total # of errors $X+Y$ also has a Poisson distribution with parameter $\lambda+\mu$.

13 a $f(x,y) = f_X(x) \cdot f_Y(y) = \begin{cases} e^{-x-y} & x \ge 0,\ y \ge 0 \\ 0 & \text{otherwise} \end{cases}$

b $P(X \le 1 \text{ and } Y \le 1) = P(X \le 1) \cdot P(Y \le 1) = (1 - e^{-1})(1 - e^{-1}) = .400$

c $P(X+Y \le 2) = \int_0^2\int_0^{2-x} e^{-x-y}\,dy\,dx = \int_0^2 e^{-x}[1 - e^{-(2-x)}]\,dx$

$$= \int_0^2 (e^{-x} - e^{-2})\,dx = 1 - e^{-2} - 2e^{-2} = .594$$

d $P(X+Y \le 1) = \int_0^1 e^{-x}[1 - e^{-(1-x)}]\,dx = 1 - 2e^{-1} = .264,$

so $P(1 \le X+Y \le 2) = P(X+Y \le 2) - P(X+Y \le 1) = .594 - .264 = .330$

15 a $F(y) = P(Y \le y) = P[(X_1 \le y) \cup ((X_2 \le y) \cap (X_3 \le y))]$

$= P(X_1 \le y) + P[(X_2 \le y) \cap (X_3 \le y)] - P[(X_1 \le y) \cap (X_2 \le y) \cap (X_3 \le y)]$

$= (1 - e^{-\lambda y}) + (1 - e^{-\lambda y})^2 - (1 - e^{-\lambda y})^3$ for $y \ge 0$

$f(y) = F'(y) = \lambda e^{-\lambda y} + 2(1 - e^{-\lambda y})(\lambda e^{-\lambda y}) - 3(1 - e^{-\lambda y})^2(\lambda e^{-\lambda y})$

$= 4\lambda e^{-2\lambda y} - 3\lambda e^{-3\lambda y}$ $\quad y \ge 0$

b $E(Y) = \int_0^\infty y \cdot (4\lambda e^{-2\lambda y} - 3\lambda e^{-3\lambda y})\,dy = 2\left(\dfrac{1}{2\lambda}\right) - \dfrac{1}{3\lambda} = \dfrac{2}{3\lambda}$

17 a $P\left((X,Y) \text{ within a circle of radius } \dfrac{R}{2}\right) = P(A) = \iint_A f(x,y)\,dx\,dy$

$$= \frac{1}{\pi R^2}\iint_A dx\,dy = \frac{\text{area of } A}{\pi R^2} = \frac{1}{4} = .25$$

b

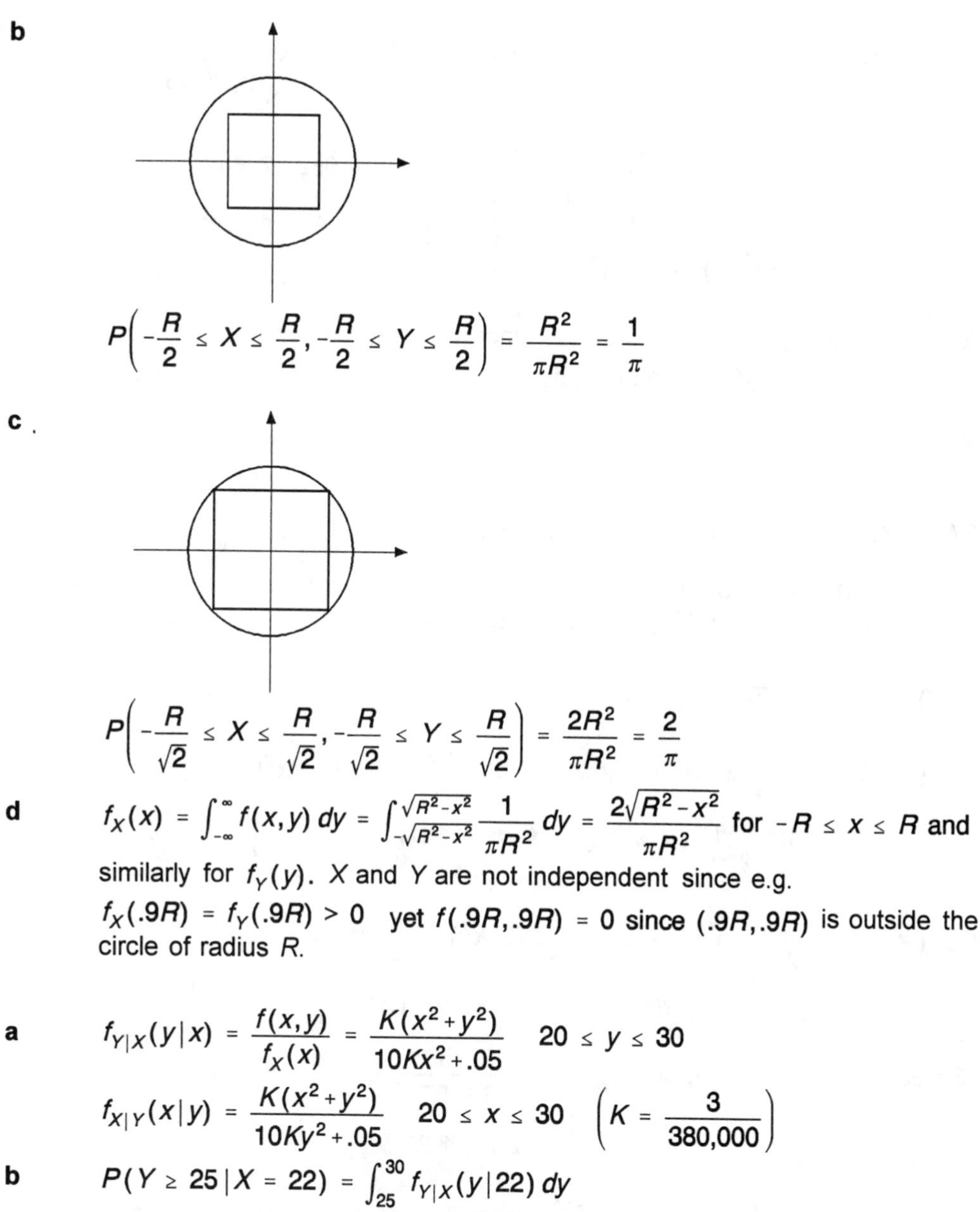

$$P\left(-\frac{R}{2} \le X \le \frac{R}{2}, -\frac{R}{2} \le Y \le \frac{R}{2}\right) = \frac{R^2}{\pi R^2} = \frac{1}{\pi}$$

c

$$P\left(-\frac{R}{\sqrt{2}} \le X \le \frac{R}{\sqrt{2}}, -\frac{R}{\sqrt{2}} \le Y \le \frac{R}{\sqrt{2}}\right) = \frac{2R^2}{\pi R^2} = \frac{2}{\pi}$$

d $f_X(x) = \int_{-\infty}^{\infty} f(x,y)\,dy = \int_{-\sqrt{R^2-x^2}}^{\sqrt{R^2-x^2}} \frac{1}{\pi R^2}\,dy = \frac{2\sqrt{R^2-x^2}}{\pi R^2}$ for $-R \le x \le R$ and similarly for $f_Y(y)$. X and Y are not independent since e.g. $f_X(.9R) = f_Y(.9R) > 0$ yet $f(.9R,.9R) = 0$ since $(.9R,.9R)$ is outside the circle of radius R.

19 **a** $f_{Y|X}(y|x) = \frac{f(x,y)}{f_X(x)} = \frac{K(x^2+y^2)}{10Kx^2+.05}$ $\quad 20 \le y \le 30$

$f_{X|Y}(x|y) = \frac{K(x^2+y^2)}{10Ky^2+.05}$ $\quad 20 \le x \le 30$ $\quad \left(K = \frac{3}{380{,}000}\right)$

b $P(Y \ge 25 \,|\, X = 22) = \int_{25}^{30} f_{Y|X}(y|22)\,dy$

$$= \int_{25}^{30} \frac{k((22)^2+y^2)}{10k(22)^2+.05}\,dy = .556$$

$P(Y \ge 25) = \int_{25}^{30} f_y(y)\,dy = \int_{25}^{30} (10ky^2+.05)\,dy = .549$

c $E(Y|X=22) = \int_{-\infty}^{\infty} y \cdot f_{Y|X}(y|22)\, dy = \int_{20}^{30} y \cdot \frac{k((22)^2+y^2)}{10k(22)^2+.05}\, dy$

$= 25.372912$

$E(Y^2|X=22) = \int_{25}^{30} y^2 \cdot \frac{k((22)^2+y^2)}{10k(22)^2+.05}\, dy = 652.028640$

$V(Y|X=22) = E(Y^2|X=22) - [E(Y|X=22)]^2 = 8.243976$

$\sigma_{Y|X=22} = [V(Y|X=22)]^{1/2} = 2.871233$

21 For every x and y, $f_{Y|X}(y|x) = f_y(y)$, since then $f(x,y) = f_{Y|X}(y|x) \cdot f_X(x) = f_Y(y) \cdot f_X(x)$ as required.

Section 5.2

23 $E(X_1 - X_2) = \sum_{x_1=0}^{4} \sum_{x_2=0}^{3} (x_1 - x_2) \cdot p(x_1, x_2)$

$= (0-0) \cdot (.08) + (0-1)(.07) + \ldots + (4-3)(.06) = .15$

(which equals $E(X_1) - E(X_2) = 1.70 - 1.55$)

25 $E(XY) = E(X) \cdot E(Y) = L \cdot L = L^2$

27 $E[h(X,Y)] = \int_0^1 \int_0^1 |x-y| \cdot 6x^2 y\, dx\, dy = 2 \int_0^1 \int_0^x (x-y) \cdot 6x^2 y\, dy\, dx$

$= 12 \int_0^1 \int_0^x (x^3 y - x^2 y^2)\, dy\, dx = 12 \int_0^1 \frac{x^5}{6}\, dx = \frac{1}{3}$

29 $\text{Cov}(X,Y) = -\frac{2}{75}$ and $\mu_X = \mu_Y = \frac{2}{5}$. $E(X^2) = \int_0^1 x^2 \cdot f_X(x)\, dx$

$= 12 \int_0^1 x^3 (1-x)^2\, dx = \frac{12}{60} = \frac{1}{5}$, so $\text{Var}(X) = \frac{1}{5} - \frac{4}{25} = \frac{1}{25}$

Similarly, $\text{Var}(Y) = \frac{1}{25}$, so $\rho_{X,Y} = \frac{-2/75}{\sqrt{\frac{1}{25}} \cdot \sqrt{\frac{1}{25}}} = -\frac{50}{75} = -.667$

31 a $E(X) = \int_{20}^{30} x\, f_X(x)\, dx = \int_{20}^{30} x[10Kx^2 + .05]\, dx = 25.328947 = E(Y)$

$E(XY) = \int_{20}^{30} \int_{20}^{30} xy \cdot K(x^2+y^2)\, dx\, dy = 641.447368$

$\Rightarrow \text{Cov}(X,Y) = 641.447368 - (25.328947)^2 = -.1082$

b $E(X^2) = \int_{20}^{30} x^2[10Kx^2 + .05]\,dx = 649.8246 = E(Y^2)$, so $\text{Var}(X)$

$= \text{Var}(Y) = 649.8246 - (25.329)^2 = 8.2664$

$$\Rightarrow \rho = \frac{-.1082}{\sqrt{(8.2664)(8.2664)}} = -.0131$$

33 Since $E(XY) = E(X) \cdot E(Y)$, $\text{Cov}(X,Y) = E(XY) - E(X) \cdot E(Y)$

$= E(X) \cdot E(Y) - E(X) \cdot E(Y) = 0$,

and since $\text{Corr}(X,Y) = \dfrac{\text{Cov}(X,Y)}{\sigma_X \sigma_Y}$ then $\text{corr}(X,Y) = 0$

35 **a** $\text{Cov}(aX+b, cY+d) = E[(aX+b)(cY+d)] - E(aX+b) \cdot E(cY+d)$

$= E[acXY + adX + bcY + bd] - (aE(X) + b)(cE(Y) + d)$

$= acE(XY) - acE(X)\,E(Y) = ac\text{Cov}(X,Y)$

b $\text{Corr}(aX+b, cY+d) = \dfrac{\text{Cov}(aX+b, cY+d)}{\sqrt{\text{Var}(aX+b)}\sqrt{\text{Var}(cY+d)}} = \dfrac{ac\,\text{Cov}(X,Y)}{|a| \cdot |c|\sqrt{\text{Var}(X) \cdot \text{Var}(Y)}}$

$= \text{Corr}(X,Y)$ when a and c have the same signs. When a and c differ in sign,

$\text{Corr}(aX+b, cY+d) = -\text{Corr}(X,Y)$.

Section 5.3

37

	$p(x_1)$	.50	.30	.20
$p(x_2)$	x_2/x_1	160	200	250
.50	160	.25	.15	.10
.30	200	.15	.09	.06
.20	250	.10	.06	.04

a

$\bar{x}$	160	180	200	205	225	250
$p(\bar{x})$	.25	.30	.09	.20	.12	.04

$E(\bar{X}) = 190 = \mu$

b

s^2	0	800	1250	4050
$p(s^2)$	.38	.30	.12	.20

$E(S^2) = 1200 = \sigma^2$

39

x	0	1	2	3	4	5	6	7	8	9	10
$\frac{x}{n}$	0	.1	.2	.3	.4	.5	.6	.7	.8	.9	1.0
$p\left(\frac{x}{n}\right)$	.000	.000	.000	.001	.005	.027	.088	.201	.302	.269	.107

X is a binomial random variable (with p = .8).

41

outcome:	1,1	1,2	1,3	1,4	2,1	2,2	2,3	2,4
probability:	.16	.12	.08	.04	.12	.09	.06	.03
$\bar{x}$:	1	1.5	2	2.5	1.5	2	2.5	3
r:	0	1	2	3	1	0	1	2

outcome:	3,1	3,2	3,3	3,4	4,1	4,2	4,3	4,4
probability:	.08	.06	.04	.02	.04	.03	.02	.01
$\bar{x}$:	2	2.5	3	3.5	2.5	3	3.5	4
r:	2	1	0	1	3	2	1	0

a

$\bar{x}$	1	1.5	2	2.5	3	3.5	4
$p(\bar{x})$	.16	.24	.25	.20	.10	.04	.01

b $P(\bar{X} \leq 2.5) = .85$

c

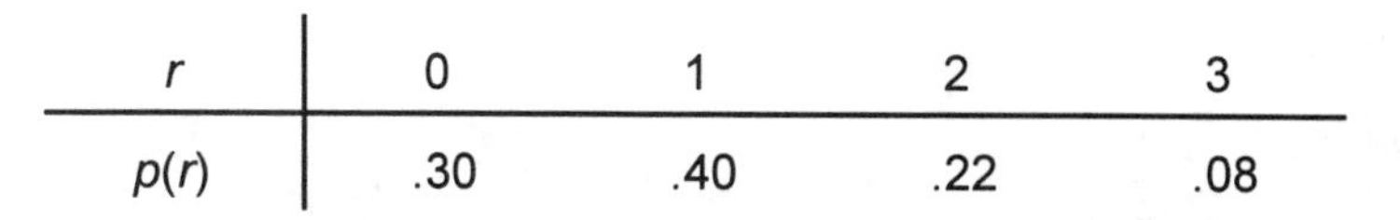

r	0	1	2	3
$p(r)$	.30	.40	.22	.08

d

$$\begin{aligned} P(\bar{X} \le 1.5) &= P(1,1,1,1) + P(2,1,1,1) + \ldots + P(1,1,1,2) + P(1,1,2,2) + \ldots \\ &\quad + P(2,2,1,1) + P(3,1,1,1) + \ldots + P(1,1,1,3) \\ &= (.4)^4 + 4(.4)^3(.3) + 6(.4)^2(.3)^2 + 4(.4)^2(.2)^2 \\ &= .2400 \end{aligned}$$

43 The statistic of interest is the fourth spread, or the difference between the medians of the upper and lower halves of the data. The population distribution is uniform with $A = 8$ and $B = 10$. Use a computer to generate samples of sizes $n = 5$, 10, 20, and 30 from a uniform distribution with $A = 8$ and $B = 10$. Keep the number of replications the same (say for example 500). For each sample compute the upper and lower fourth, then compute the difference. Plot the sampling distributions on separate histograms for $n = 5$, 10, 20, and 30.

45 $\mu = 12$ cm $\quad \sigma = .04$ cm

a $n = 16$

$$\begin{aligned} P(11.99 \le \bar{X} \le 12.01) &= P\left(\frac{11.99 - 12}{.01} \le Z \le \frac{12.01 - 12}{.01}\right) \\ &= P(-1 \le Z \le 1) \\ &= \Phi(1) - \Phi(-1) \\ &= .8413 - .1587 \\ &= .6826 \end{aligned}$$

b $n = 25$

$$\begin{aligned} P(\bar{X} > 12.01) = P\left(Z > \frac{12.01 - 12}{.04/5}\right) &= P(Z > 1.25) \\ &= 1 - \Phi(1.25) \\ &= 1 - .8944 \\ &= .1056 \end{aligned}$$

47 **a** 11 P.M. – 6:50 P.M. = 250 min. With $T_o = X_1 + \ldots + X_{40}$ = total grading time, $\mu_{T_o} = n\mu = (40)(6) = 240$ and $\sigma_{T_o} = \sigma\sqrt{n} = 37.95$, so $P(T_0 \le 250)$

$$\approx P\left(Z \le \frac{250 - 240}{37.95}\right) = P(Z \le .26) = .6026$$

b $$P(T_0 > 260) \approx P\left(Z > \frac{260 - 240}{37.95}\right) = P(Z > .53) = .2981$$

49 $x \sim N(10,4)$

for day 1, $n = 5$

$$P(\bar{X} \le 11) = P\left(Z \le \frac{11-10}{2/\sqrt{5}}\right)$$
$$= P(Z \le 1.12)$$
$$= .8686$$

for day 2, $n = 6$

$$P(\bar{X} \le 11) = P\left(Z \le \frac{11-10}{2/\sqrt{6}}\right)$$
$$= P(Z \le 1.22)$$
$$= .8888$$

for both days

$$P(\bar{X} \le 11) = (.8686)(.8888)$$
$$= .7720$$

51 $\mu = 50, \quad \sigma = 1.5$

a $n = 9$

$$P(X \ge 52) = P\left(Z > \frac{52-50}{1.5/3}\right)$$
$$= P(Z > 4.0)$$
$$= 1 - \Phi(4.0) = 0$$

b $n = 40$

$$P(X \ge 52) = P\left(Z > \frac{52-50}{1.5/\sqrt{40}}\right)$$
$$= P(Z > 8.43)$$
$$= 1 - \Phi(8.43) = 0$$

53 $\mu = np = 20, \ \sigma = \sqrt{npq} = 3.464$

a $P(25 \le X) \approx P\left(\frac{24.5-20}{3.464} \le Z\right) = P(1.30 \le Z) = .0968$

b $P(15 \le X \le 25) \approx P\left(\frac{14.5-20}{3.464} \le Z \le \frac{25.5-20}{3.464}\right)$

$$= P(-1.59 \le Z \le 1.59) = .8882$$

55 $E(X) = 100$, $\text{Var}(X) = 200$, $\sigma_X = 14.14$, so $P(X \le 125) \approx P\left(Z \le \frac{125-100}{14.14}\right)$

$$= P(Z \le 1.77) = .9616$$

Chapter 5

Section 5.4

57

a $X_1+X_2+X_3$ has a normal distribution with mean $\mu_1+\mu_2+\mu_3 = 300$ and variance $\sigma_1^2+\sigma_2^2+\sigma_3^2 = 36$, so $P(X_1+X_2+X_3 \le 309) = P\left(Z \le \frac{[309-300]}{6}\right)$

$= P(Z \le 1.5) = .9332$, and $P(288 \le X_1+X_2+X_3 < 312)$

$= P\left(\frac{[288-300]}{6} \le Z \le \frac{[312-300]}{6}\right)$

$= P(-2 \le Z \le 2) = .9544.$

b $\mu_{\bar{X}} = 100,\ \sigma_{\bar{X}} = \frac{\sigma}{\sqrt{n}} = \frac{\sqrt{12}}{\sqrt{3}} = \sqrt{4} = 2$, so $P(105 \le \bar{X})$

$= P\left(\frac{105-100}{2} \le Z\right) = P(2.5 \le Z) = .0062$, and $P(98 \le \bar{X} \le 102)$

$= P(-1 \le Z \le 1) = .6826$

c $X_1-.5X_2-.5X_3$ is normally distributed with mean $100-.5(100)-.5(100) = 0$ and variance $12+(.5)^2(12)+(.5)^2(12) = 18$, so $P(-10 \le X_1-.5X_2-.5X_3 \le 5)$

$= P\left(\frac{-10}{\sqrt{18}} \le Z \le \frac{5}{\sqrt{18}}\right) = .8719.$

d $X_1+X_2+X_3$ is normal with mean $90+100+110 = 300$ and variance $10+12+14 = 36$, so $P(X_1+X_2+X_3 \le 306) = P(Z \le 1) = .8413$, and

$P\left(98 \le \frac{X_1+X_2+X_3}{3} \le 102\right) = P(294 \le X_1+X_2+X_3 \le 306) = 0.6826.$

59

a The marginal pmf's of X and Y are given in the solution to Exercise 7, from which $E(X) = 2.8,\ E(Y) = .7,\ V(X) = 1.66,\ V(Y) = .61$. Thus $E(X+Y) = E(X)+E(Y) = 3.5,\ V(X+Y) = V(X)+V(Y) = 2.27$, and the standard deviation of $X+Y$ is 1.51.

b $E(3X+10Y) = 3E(X)+10E(Y) = 15.4,\ V(3X+10Y)$
$= 9V(X)+100V(Y) = 75.94$, and the standard deviation of revenue is 8.71.

61

a $E(X_1) = 1.70,\ E(X_2) = 1.55,\ E(X_1X_2) = \sum_{x_1}\sum_{x_2} x_1x_2\,p(x_1,x_2) = 3.33$

so $\text{Cov}(X_1,X_2) = E(X_1X_2)-E(X_1)E(X_2) = 3.33-2.635 = .695$

b $V(X_1+X_2) = V(X_1)+V(X_2)+2\text{Cov}(X_1,X_2)$

$= 1.59+1.0875+2(.695) = 4.0675$

Chapter 5

63 **a** $\mu_{\bar{X}-\bar{Y}} = \mu_{\bar{X}} - \mu_{\bar{Y}} = 5.00 - 5.00 = 0$, $\sigma^2_{\bar{X}-\bar{Y}} = \sigma^2_{\bar{X}} + \sigma^2_{\bar{Y}} = \frac{.09}{25} + \frac{.09}{25}$

$= \frac{.18}{25}$, $\sigma_{\bar{X}-\bar{Y}} = .08485$. Thus, $P(-.1 \le \bar{X} - \bar{Y} \le .1)$

$= P\left(\frac{-.1}{.08485} \le Z \le \frac{0.1}{.08485}\right) = P(-1.18 \le Z \le 1.18) = .7620$

b $\sigma^2_{\bar{X}-\bar{Y}} = \frac{.09}{36} + \frac{.09}{36} = .005$, so $\sigma_{\bar{X}-\bar{Y}} = .0707$ and $P(-.1 \le \bar{X} - \bar{Y} \le .1)$

$\doteq P(-1.41 \le Z \le 1.41) = .8414$

65 Letting X_1, X_2, and X_3 denote the lengths of the three pieces, the total length is $X_1 + X_2 - X_3$. This has a normal distribution with mean value $20 + 15 - 1 = 34$, variance $.25 + .16 + .01 = .42$, and standard deviation 0.6481. Standardizing gives $P(34.5 \le X_1 + X_2 - X_3 \le 35) = P(.77 \le Z \le 1.54) = .1588$

67 **a** $E(X_1 + X_2 + X_3) = 800 + 1000 + 600 = 2400$

b Assuming independence of

X_1, X_2, X_3, $\text{Var}(X_1 + X_2 + X_3) = (16)^2 + (25)^2 + (18)^2 = 1205$

c $E(X_1 + X_2 + X_3) = 2400$ as before, but now $\text{Var}(X_1 + X_2 + X_3) = \text{Var}(X_1)$

$+ \text{Var}(X_2) + \text{Var}(X_3) + 2\,\text{Cov}(X_1, X_2) + 2\,\text{Cov}(X_1, X_3)$

$+ 2\,\text{Cov}(X_2, X_3) = 1745$, $\sigma_{X_1+X_2+X_3} = 41.77$

69 **a** $M = a_1X_1 + a_2X_2 + W\int_0^{12} x\,dx = a_1X_1 + a_2X_2 + 72W$, so

$E(M) = (5)(2) + (10)(4) + (72)(1.5) = 158$,

$\sigma^2_M = (5)^2(.5)^2 + (10)^2(1)^2 + (72)^2(.25)^2 = 430.25$,

$\sigma_M = 20.74$

b $P(M \le 200) = P\left(Z \le \frac{200 - 158}{20.74}\right) = P(Z \le 2.03) = .9788$

71 **a** Both approximately normal by the CLT

b The difference of two rv's is just a special linear combination, and a linear combination of normal rv's has a normal distribution, so $\bar{X} - \bar{Y}$ has approximately a normal distribution with $\mu_{\bar{X}-\bar{Y}} = 2$ and

$\sigma^2_{\bar{X}-\bar{Y}} = \frac{(8)^2}{40} + \frac{(6)^2}{35} = 2.629$, $\sigma_{\bar{X}-\bar{Y}} = 1.621$

c $P(-1 \le \bar{X}-\bar{Y} \le 1) \approx P\left(\frac{-3}{1.621} \le Z \le \frac{-1}{1.621}\right)$
$= P(-1.85 \le Z \le -.62) = .2354$

d $P(\bar{X}-\bar{Y} \ge 6) \approx P(Z \ge 2.47) = 0.0068$; this probability is quite small, so such an occurrence is unlikely if $\mu_1 - \mu_2 = 2$, and we would thus doubt this claim.

Supplementary

73 a $p_X(x)$ is obtained by adding joint probabilities across the row labeled x, resulting in $p_X(x) = .2, .5, .3$ for $x = 7, 9, 10$ respectively. Similarly, from column sums $p_Y(y) = .1, .35, .55$ for $y = 7, 9, 10$ respectively.

b $P(X \le 9 \text{ and } Y \le 9) = p(7,7) + p(7,9) + p(9,7) + p(9,9) = 0.25$

c $p_X(7) \cdot p_Y(7) = (.2)(.1) \ne .05 = p(7,7)$, so X and Y are not independent (almost any other (x, y) pair yields the same conclusion).

d $E(X+Y) = \Sigma\Sigma(x+y)p(x,y)$ (or $E(X)+E(Y)$) $= 18.25$

e $E(|X-Y|) = \Sigma\Sigma\,|x-y|\,p(x,y) = 1.05$

75

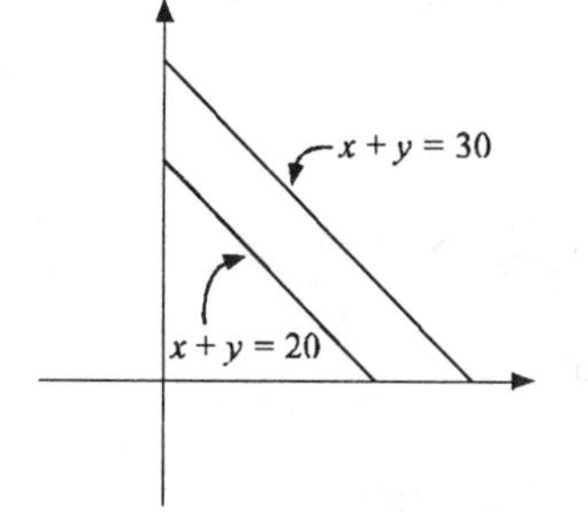

a $1 = \int_{-\infty}^{\infty}\int_{-\infty}^{\infty} f(x,y)\,dx\,dy = \int_0^{20}\int_{20-x}^{30-x} kxy\,dy\,dx + \int_{20}^{30}\int_0^{30-x} kxy\,dy\,dx$
$= \frac{81{,}250}{3} \cdot k \Rightarrow k = \frac{3}{81{,}250}$

b $$f_X(x) = \begin{cases} \int_{20-x}^{30-x} kxy\,dy = k(250x - 10x^2) & 0 \le x \le 20 \\ \int_0^{30-x} kxy\,dy = k(450x - 30x^2 + \frac{1}{2}x^3) & 20 \le x \le 30 \end{cases}$$

and by symmetry $f_Y(y)$ is obtained by substituting y for x in $f_X(x)$.
Since $f_X(25) > 0$, $f_Y(25) > 0$, but $f(25,25) = 0$, $f_X(x) \cdot f_Y(y) \ne f(x,y)$ for all x,y so X and Y are not independent.

c $$P(X+Y \le 25) = \int_0^{20}\int_{20-x}^{25-x} kxy\,dy + \int_{20}^{25}\int_0^{25-x} kxy\,dx$$
$$= \frac{3}{81{,}250} \cdot \frac{230{,}625}{24} = .355$$

d $$E(X+Y) = E(X)+E(Y) = 2\left\{\int_0^{20} x \cdot k(250x-10x^2)dx + \int_{20}^{30} x \cdot k\left(450x-30x^2+\frac{1}{2}x^3\right)dx\right\}$$
$$= 2K\{351{,}666.67\} = 25.969$$

e $$E(XY) = \int_{-\infty}^{\infty}\int_{-\infty}^{\infty} xy \cdot f(x,y)\,dx\,dy = \int_0^{20}\int_{20-x}^{30-x} kx^2y^2\,dy\,dx + \int_{20}^{30}\int_0^{30-x} kx^2y^2\,dy\,dx = \frac{K}{3} \cdot \frac{33{,}250{,}000}{3} = 136.4103, \text{ so}$$
$$\text{Cov}(X,Y) = 136.4103-(12.9845)^2 = -32.19,\ E(X^2) = E(Y^2) = 204.6154, \text{ so } \sigma_X^2 = \sigma_Y^2 = 204.6154-(12.9845)^2$$
$$= 36.0182 \text{ and } \rho = \frac{-32.19}{36.0182} = -.894$$

f $$\text{Var}(X+Y) = \text{Var}(X)+\text{Var}(Y)+2\,\text{Cov}(X,Y) = 7.66$$

77 Let $W_i = X_i+Y_i+Z_i$ = total intake on day i. Then
$E(W_i) = 500+800+1700 = 3000$ and $\text{Var}(W_i)$
$= (50)^2+(100)^2+(200)^2 = 52{,}500$ so $\sigma_{W_i} = 229.13$.
By the CLT, $\sum_{i=1}^{365} \frac{W_i}{365} = \bar{W}$ is approximately normal with
$\mu_{\bar{W}} = 3000$ and $\sigma_{\bar{W}} = \frac{229.13}{\sqrt{365}} = 11.99$, so $P(2950 \le \bar{W} \le 3050)$
$\approx P(-4.17 \le Z \le 4.17) = 1$.

79
a $E(N) \cdot \mu = (10)(40) = 400$ min.
b We expect 20 components to come in for repair during a 4 hour period, so
$E(N) \cdot \mu = (20)(3.5) = 70$

81 $$0.95 = P(\mu-0.02 \le \bar{X} \le \mu+0.02) \doteq P\left(\frac{-0.02}{0.1/\sqrt{n}} \le Z \le \frac{0.02}{0.1/\sqrt{n}}\right)$$
$= P(-0.2\sqrt{n} \le Z \le 0.2\sqrt{n})$, but $P(-1.96 \le Z \le 1.96)$
$= 0.95$, so $0.2\sqrt{n} = 1.96 \Rightarrow n = 97$. The CLT justifies this calculation.

Chapter 5

83 The expected value and standard deviation of volume are 87,850 and 4370.37, respectively, so

$$P(\text{Volume} \le 100{,}000) = P\left(Z \le \frac{100{,}000 - 87{,}850}{4370.37}\right) = P(Z \le 2.78) = 0.9973$$

85 **a** $\text{Var}(aX+Y) = a^2\sigma_X^2 + 2a\,\text{Cov}(X,Y) + \sigma_Y^2 = a^2\sigma_X^2 + 2a\sigma_X\sigma_Y\rho + \sigma_Y^2.$

Substituting $a = \dfrac{\sigma_Y}{\sigma_X}$ yields $\sigma_Y^2 + 2\sigma_Y^2\rho + \sigma_Y^2 = 2\sigma_Y^2(1+\rho) \ge 0$, so $\rho \ge -1$.

b Same argument as in (a).

c Suppose $\rho = 1$. Then $\text{Var}(aX-Y) = 2\sigma_Y^2(1-\rho) = 0$, which implies that $aX - Y = k$ (a constant), so $Y = aX - k$ which is of the form $aX + b$.

87 **a** With $Y = X_1 + X_2$, $F_Y(y) = \displaystyle\int_0^y \left\{\int_0^{y-x_1} \frac{1}{2^{\nu_1/2}\Gamma(\nu_1/2)} \cdot \frac{1}{2^{\nu_2/2}\Gamma(\nu_2/2)} \cdot x_1^{\frac{\nu_1}{2}-1} x_2^{\frac{\nu_2}{2}-1} e^{-\frac{x_1+x_2}{2}} dx_2\right\} dx_1.$ But the inner integral can be shown to equal $\dfrac{1}{2^{(\nu_1+\nu_2)/2}\Gamma((\nu_1+\nu_2)/2)} y^{[\nu_1+\nu_2]/2-1} e^{-y/2}$, from which the result follows.

b By (a), $Z_1^2 + Z_2^2$ is chi-squared with $\nu = 2$, so $(Z_1^2 + Z_2^2) + Z_3^2$ is chi-squared with $\nu = 3$, etc., until $Z_1^2 + \ldots + Z_n^2$ is chi-squared with $\nu = n$.

c $\dfrac{X_i - \mu}{\sigma}$ is standard normal, so $\left[\dfrac{X_i - \mu}{\sigma}\right]^2$ is chi-squared with $\nu = 1$, so the sum is chi-squared with $\nu = n$.

89 **a** $V(X_1) = V(W+E_1) = \sigma_W^2 + \sigma_E^2 = V(W+E_2) = V(X_2)$, and $\text{Cov}(X_1, X_2)$

$$= \text{Cov}(W+E_1, W+E_2) = \text{Cov}(W,W) + \text{Cov}(W,E_2) + \text{Cov}(E_1,W) + \text{Cov}(E_1,E_2) = \text{Cov}(W,W) = V(W) = \sigma_W^2$$

Thus, $\rho = \dfrac{\sigma_W^2}{\sqrt{\sigma_W^2 + \sigma_E^2} \cdot \sqrt{\sigma_W^2 + \sigma_E^2}} = \dfrac{\sigma_W^2}{(\sigma_W^2 + \sigma_E^2)}$

b $\rho = \dfrac{1}{1 + .0001} = .9999$

Chapter 5

91 $E(Y) \doteq h(\mu_1, \mu_2, \mu_3, \mu_4) = 120\left[\frac{1}{10} + \frac{1}{15} + \frac{1}{20}\right] = 26$

The partial derivatives of $h(x_1, x_2, x_3, x_4)$ with respect to x_1, x_2, x_3 and x_4 are $-\frac{x_4}{x_1^2}$, $-\frac{x_4}{x_2^2}$, $-\frac{x_4}{x_3^2}$ and $\frac{1}{x_1} + \frac{1}{x_2} + \frac{1}{x_3}$, respectively. Substituting $x_1 = 10$, $x_2 = 15$, $x_3 = 20$, and $x_4 = 120$ gives -1.2, -.5333, -.3000, and .2167, respectively, so $V(Y) = (1)(-1.2)^2 + (1)(-.5333)^2 + (1.5)^2(-.3000)^2 + (4.0)^2(.2167)^2 = 2.6783$, and the approximate sd of y is 1.64.

CHAPTER 6

Section 6.1

1 **a** $\Sigma x_i = 657.2$, so $\hat{\mu} = \bar{x} = \frac{657.2}{10} = 65.72$. $\Sigma x_i^2 = 43{,}376.02$, so

$s^2 = \frac{43{,}376.02-(657.2)^2/10}{9} = 20.537$, $\hat{\sigma} = s = 4.53$

b The ordered values are 57.2, 62.1, 62.5, 64.3, 64.9, 65.3, 68.7, 69.8, 70.8, 71.6, so $\tilde{X} = \frac{64.9+65.3}{2} = 65.10$; $\bar{x}$ cannot be used as an estimate of $\tilde{\mu}$ because there is no assurance that $\tilde{\mu} = \mu$.

c $\hat{\mu} = \bar{x}_{tr(10)} = \frac{62.1+62.5+...+69.8+70.8}{8} = 66.05$

d There are 7 x_i's, for which $60 \le x_i \le 70$, so $\hat{p} = \frac{7}{10} = .70$.

3 **a** $\hat{\mu} = \bar{x} = \frac{921.82}{12} = 76.818$, using the sample mean $\bar{X}$ as the estimator.

b With $\tilde{X}$ as the estimator of $\tilde{\mu}$, $\tilde{x} = \frac{76.88+76.88}{2} = 76.88$.

c With p = true proportion of samples with content at most 76% and X = # in samples that are ≤ 76, $\hat{p} = \frac{X}{n}$; here $x = 1$, so $\hat{p} = \frac{1}{12} = .083$.

d The standard error of $\bar{X}$ is $\frac{\sigma}{\sqrt{n}}$, so the estimated standard error is

$\frac{s}{\sqrt{n}} = \frac{.748}{\sqrt{12}} = .216$.

5 **a** $E(\bar{X}-\bar{Y}) = E(\bar{X}) - E(\bar{Y}) = \mu_1 - \mu_2$ as desired. With $\bar{x} = 76.82$ and $\bar{y} = 74.28$, $(\widehat{\mu_1 - \mu_2}) = \bar{x}-\bar{y} = 2.54$.

b $\text{Var}(\bar{X}-\bar{Y}) = \text{Var}(\bar{X}) + \text{Var}(\bar{Y}) = \frac{\sigma_1^2}{m}+\frac{\sigma_2^2}{n}$, and $\sigma_{\bar{X}-\bar{Y}} = \left(\frac{\sigma_1^2}{m}+\frac{\sigma_2^2}{n}\right)^{1/2}$

c Estimated standard error of $\bar{X}-\bar{Y} = \left(\frac{s_1^2}{m}+\frac{s_2^2}{n}\right)^{1/2} = \left(\frac{.5625}{12}+\frac{.4900}{14}\right)^{1/2} = .286$.

Chapter 6

7 **a** $\hat{\mu} = \bar{x} = \frac{\Sigma x_i}{n} = \frac{1206}{10} = 120.6$

b $\hat{\tau} = 10{,}000\hat{\mu} = 1{,}206{,}000$

c 8 of 10 house in the sample used at least 100 therms (the "successes"), so $\hat{p} = \frac{8}{10} = .80.$

d The ordered sample values are 89, 99, 103, 109, 118, 122, 125, 138, 147, 156, from which the two middle values are 118 and 122, so

$$\hat{\tilde{\mu}} = \tilde{x} = \frac{118 + 122}{2} = 120.0$$

9 **a** $E(\bar{X}) = \mu = E(X) = \lambda$, so $\bar{X}$ is an unbiased estimator for the Poisson parameter λ; $\Sigma x_i = (0)(18) + (1)(37) + \ldots + (7)(1) = 317.$, since n = 150, $\hat{\lambda} = \bar{x} = \frac{317}{150} = 2.11.$

b $\sigma_{\bar{X}} = \frac{\sigma}{\sqrt{n}} = \frac{\sqrt{\lambda}}{\sqrt{n}}$, so the estimated standard error is $\sqrt{\frac{\hat{\lambda}}{\sqrt{n}}} = \frac{\sqrt{2.11}}{\sqrt{150}} = 0.119.$

11 **a** $E\left(\frac{X_1}{n_1} - \frac{X_2}{n_2}\right) = \frac{1}{n_1}E(X_1) - \frac{1}{n_2}E(X_2) = \frac{1}{n_1}(n_1p_1) - \frac{1}{n_2}(n_2p_2) = p_1 - p_2$

b $\text{Var}\left(\frac{X_1}{n_1} - \frac{X_2}{n_2}\right) = \text{Var}\left(\frac{X_1}{n_1}\right) + \text{Var}\left(\frac{X_2}{n_2}\right) = \left(\frac{1}{n_1}\right)^2\text{Var}(X_1) + \left(\frac{1}{n_2}\right)^2\text{Var}(X_2) =$ $\frac{1}{n_1^2}(n_1p_1q_1) + \frac{1}{n_2^2}(n_2p_2q_2) = \frac{p_1q_1}{n_1} + \frac{p_2q_2}{n_2}$, and the standard error is the square root of this quantity.

c With $\hat{p}_1 = \frac{x_1}{n_1}$, $\hat{q}_1 = 1 - \hat{p}_1$, $\hat{p}_2 = \frac{x_2}{n_2}$, $\hat{q}_2 = 1 - \hat{p}_2$, the estimated standard error is $\left(\frac{\hat{p}_1\hat{q}_1}{n_1} + \frac{\hat{p}_2\hat{q}_2}{n_2}\right)^{1/2}$.

d $(\hat{p}_1 - \hat{p}_2) = \frac{127}{200} - \frac{176}{200} = 0.635 - 0.880 = -0.245.$

e $\left(\frac{(0.635)(0.365)}{200} + \frac{(0.880)(0.120)}{200}\right)^{1/2} = 0.041.$

Chapter 6

13 $E(X) = \int_{-1}^{1} x \cdot \frac{1}{2}(1+\theta x) = \frac{x^2}{4} + \frac{\theta x^3}{6}\Big|_{-1}^{1} = \frac{1}{3}\theta$

$E(X) = \frac{1}{3}\theta \qquad E(\bar{X}) = \frac{1}{3}\theta$

$\hat{\theta} = 3\bar{X} \Rightarrow E(\hat{\theta}) = E(3\bar{X}) = 3E(\bar{X}) = 3\left(\frac{1}{3}\theta\right) = \theta$

15 **a** $E(X^2) = 2\theta$ implies that $E(X^2/2) = \theta$. Consider $\hat{\theta} = \frac{\Sigma X_i^2}{2n}$. Then

$E(\hat{\theta}) = E\left[\frac{\Sigma X_i^2}{2n}\right] = \frac{\Sigma E(X_i^2)}{2n} = \frac{\Sigma 2\theta}{2n} = \frac{2n\theta}{2n} = \theta$, implying that $\hat{\theta}$ is an unbiased estimator for θ.

b $\Sigma x_i^2 = 1490.1058$, so $\hat{\theta} = \frac{1490.1058}{20} = 74.505$

17 **a** $E(\hat{p}) = \sum_{x=0}^{\infty} \frac{r-1}{x+r-1} \cdot \binom{x+r-1}{x} \cdot p^r \cdot (1-p)^x$

$= p \sum_{x=0}^{\infty} \frac{(x+r-2)!}{x!(r-2)!} p^{r-1}(1-p)^x = p \sum_{x=0}^{\infty} \binom{x+r-2}{x} p^{r-1}(1-p)^x$

$= p \sum_{x=0}^{\infty} nb(x;\, r-1,\, p) = p.$

b For the given sequence, $x = 5$, so $\hat{p} = \frac{5-1}{5+5-1} = \frac{4}{9} = .444.$

19 **a** $\lambda = 0.5p + 0.15 \Rightarrow 2\lambda = p + 0.3$, so $p = 2\lambda - 0.3$ and $\hat{p} = 2\hat{\lambda} - 0.3 = 2\left(\frac{Y}{n}\right) - 0.3$;

the estimate is $2\left(\frac{20}{80}\right) - 0.3 = 0.2$.

b $E(\hat{p}) = E(2\hat{\lambda} - 0.3) = 2E(\hat{\lambda}) - 0.3 = 2\lambda - 0.3 = p$ as desired.

c Here $\lambda = 0.7p + (0.3)(0.3)$, so $p = \frac{10}{7}\lambda - \frac{9}{70}$ and $\hat{p} = \frac{10}{7}\left(\frac{Y}{n}\right) - \frac{9}{70}$.

Chapter 6

Section 6.2

21 **a** $E(X) = \beta \cdot \Gamma\left(1 + \frac{1}{\alpha}\right)$ and $E(X^2) = \text{Var}(X) + [E(X)]^2 = \beta^2\Gamma\left(1 + \frac{2}{\alpha}\right)$, so the moment estimators $\hat{\alpha}$ and $\hat{\beta}$ are the solution to $\bar{X} = \hat{\beta} \cdot \Gamma\left(1 + \frac{1}{\hat{\alpha}}\right)$, $\frac{1}{n}\Sigma X_i^2 = \hat{\beta}^2\Gamma\left(1 + \frac{2}{\hat{\alpha}}\right)$

Thus $\hat{\beta} = \frac{\bar{X}}{\Gamma\left(1 + \frac{1}{\hat{\alpha}}\right)}$, so once $\hat{\alpha}$ has been determined $\Gamma\left(1 + \frac{1}{\hat{\alpha}}\right)$. is evaluated and $\hat{\beta}$ then computed. Since $\bar{X}^2 = \hat{\beta}^2\Gamma^2\left(1 + \frac{1}{\hat{\alpha}}\right)$, $\frac{1}{n}\Sigma\frac{X_i^2}{\bar{X}^2} = \Gamma\left(1 + \frac{2}{\hat{\alpha}}\right)/\Gamma^2\left(1 + \frac{1}{\hat{\alpha}}\right)$, so this equation must be solved to obtain $\hat{\alpha}$.

b From (a), $\frac{1}{20}(16{,}500)/(28.0)^2 = 1.05 = \Gamma(1 + 2/\hat{\alpha})/\Gamma^2(1 + \frac{1}{\hat{\alpha}})$, so $\frac{1}{1.05} = 0.95 = \Gamma^2(1 + 1/\hat{\alpha})/\Gamma(1 + 2/\hat{\alpha})$, and from the hint, $\frac{1}{\hat{\alpha}} = 0.2 \Rightarrow \hat{\alpha} = 0.5$.
Then $\hat{\beta} = \frac{\bar{x}}{\Gamma(1.2)} = \frac{28.0}{\Gamma(1.2)}$.

23 For a single sample from a Poisson distribution,
$f(x_1,\ldots,x_n;\ \lambda) = \frac{e^{-\lambda}\lambda^{x_1}}{x_1!}\cdots\frac{e^{-\lambda}\lambda^{x_n}}{x_n!} = \frac{e^{-n\lambda}\lambda^{\Sigma x_i}}{x_1!\ldots x_n!}$, so
$\ln[f(x_1,\ldots,x_n;\ \lambda)] = -n\lambda + \Sigma x_i \ln(\lambda) - \Sigma\ln(x_i\,!)$. Thus
$\frac{d}{d\lambda}[\ln[f(x_1,\ldots,x_n;\ \lambda)] = -n + \frac{\Sigma x_i}{\lambda} = 0 \Rightarrow \hat{\lambda} = \frac{\Sigma x_i}{n} = \bar{x}$. For our problem,
$f(x_1,\ldots,x_m,y_1,\ldots,y_n;\ \lambda_1,\ \lambda_2)$ is a product of the x sample likelihood and the y sample likelihood, implying that $\hat{\lambda}_1 = \bar{x}$, $\hat{\lambda}_2 = \bar{y}$, and (by the invariance principle)
$(\widehat{\lambda_1 - \lambda_2}) = \bar{x} - \bar{y}$.

25 **a** $\hat{\mu} = \bar{x} = 384.4$; $s^2 = 395.16$, so $\frac{1}{n}\Sigma(x_i - \bar{x})^2 = \hat{\sigma}^2 = \frac{9}{10}(395.16) = 355.64$ and $\hat{\sigma} = \sqrt{355.64} = 18.86$ (this is *not* s).

b The 95th percentile is $\mu + 1.645\sigma$, so the mle of this is (by the invariance principle) $\hat{\mu} + 1.645\hat{\sigma} = 415.42$.

Chapter 6

27 **a** $f(x_1,\ldots,x_n;\ \alpha,\ \beta) = (x_1x_2\ldots x_n)^{\alpha-1}e^{-\Sigma x_i/\beta}/\beta^{n\alpha}\Gamma^n(\alpha)$, so the log likelihood is $(\alpha-1)\Sigma\ln(x_i) - \Sigma x_i/\beta - n\alpha\ln(\beta) - n\ln\Gamma(\alpha)$. Equating both $\frac{d}{d\alpha}$ and $\frac{d}{d\beta}$ to 0 yields $\Sigma\ln(x_i) - n\ln(\beta) - n\frac{d}{d\alpha}\Gamma(\alpha) = 0$, $\frac{\Sigma x_i}{\beta^2} - \frac{n\alpha}{\beta} = 0$, a very difficult system of equations to solve.

b From the second equation in (a), $\frac{\Sigma x_i}{\beta} = n\alpha \Rightarrow \bar{x} = \alpha\beta = \mu$, so the mle of μ is $\hat{\mu} = \bar{X}$.

29 **a** The joint pdf (likelihood function) is

$$f(x_1\ldots,x_n;\ \lambda,\ \theta) = \begin{cases}\lambda^n e^{-\lambda\Sigma(x_i-\theta)} & x_1 \geq \theta,\ldots,x_n \geq \theta \\ 0 & \text{otherwise}\end{cases}$$

Notice that $x_1 \geq \theta,\ldots,x_n \geq \theta$ iff $\min(x_i) \geq \theta$, and that $-\lambda\Sigma(x_i-\theta) = -\lambda\Sigma x_i + n\lambda\theta$.

$$\text{Thus likelihood} = \begin{cases}\lambda^n\exp[-\lambda\Sigma x_i]\exp[n\lambda\theta] & \min(x_i) \geq \theta \\ 0 & \min(x_i) < \theta\end{cases}$$

Consider maximization wrt θ. Because the exponent $n\lambda\theta$ is positive, increasing θ will increase the likelihood *provided that* $\min(x_i) \geq \theta$; if we make θ larger than $\min(x_i)$, the likelihood drops to 0. This implies that the mle of θ is $\hat{\theta} = \min(x_i)$. The log likelihood is now $n\ln(\lambda) - \lambda\Sigma(x_i-\hat{\theta})$. Equating the derivative wrt λ to 0 and solving yields $\hat{\lambda} = \frac{n}{\Sigma(x_i-\hat{\theta})} = \frac{n}{\Sigma x_i - n\hat{\theta}}$.

b $\hat{\theta} = \min(x_i) = 0.64$, and $\Sigma x_i = 55.80$, so $\hat{\lambda} = \frac{10}{55.80-6.4} = .202$.

Supplementary Exercises

31 $P(|\bar{X}-\mu| > \varepsilon) = P(\bar{X}-\mu > \varepsilon) + P(\bar{X}-\mu < -\varepsilon) = P\left(\frac{\bar{X}-\mu}{\sigma/\sqrt{n}} > \frac{\varepsilon}{\sigma/\sqrt{n}}\right) + P\left(\frac{\bar{X}-\mu}{\sigma/\sqrt{n}} < -\frac{\varepsilon}{\sigma/\sqrt{n}}\right)$

$= P\left(Z > \frac{\sqrt{n}\varepsilon}{\sigma}\right) + P\left(Z < \frac{-\sqrt{n}\varepsilon}{\sigma}\right) = \int_{\sqrt{n}\varepsilon/\sigma}^{\infty}\frac{1}{\sqrt{2\pi}}e^{-z^2/2}\,dz + \int_{-\infty}^{-\sqrt{n}\varepsilon/\sigma}\frac{1}{\sqrt{2\pi}}e^{-z^2/2}\,dz.$

As $n \to \infty$, both integrals $\to 0$ since $\lim_{c\to\infty}\int_c^{\infty}\frac{1}{\sqrt{2\pi}}e^{-z^2/2}\,dz = 0$.

Chapter 6

33 Let x_1 = the time until the first birth, x_2 = the elapsed time between the first and second births, and so on. Then $f(x_1,...,x_n;\ \lambda) = \lambda e^{-\lambda x_1} \cdot (2\lambda)e^{-2\lambda x_2}...(n\lambda)e^{-n\lambda x_n} = n!\lambda^n e^{-\lambda\Sigma kx_k}$. Thus the log likelihood is $\ln(n!) + n\ln(\lambda) - \lambda\Sigma k x_k$. Taking $\frac{d}{d\lambda}$ and equating to 0 yields $\hat{\lambda} = \frac{n}{\sum_{k=1}^{n} kx_k}$. For the given sample, n = 6, x_1 = 25.2, x_2 = 41.7-25.2 = 16.5, x_3 = 9.5, x_4 = 4.3, x_5 = 4.0, x_6 = 2.3; so $\sum_{k=1}^{6} kx_k = (1)(25.2) + (2)(16.5) + ... + (6)(2.3) = 137.7$ and $\hat{\lambda} = \frac{6}{137.7} = .0436$.

35

$(x_i + x_j)/2$	23.5	26.3	28.0	28.2	29.4	29.5	30.6	31.6	33.9	49.3
23.5	23.5	24.9	25.75	25.85	26.45	26.5	27.05	27.55	28.7	36.4
26.3		26.3	27.15	27.25	27.85	27.9	28.45	28.95	30.1	37.8
28.0			28.0	28.1	28.7	28.75	29.3	29.8	30.95	38.65
28.2				28.2	28.8	28.85	29.4	29.9	31.05	38.75
29.4					29.4	29.45	30	30.5	30.65	39.35
29.5						29.5	30.05	30.55	31.7	39.4
30.6							30.6	31.1	32.25	39.95
31.6								31.6	32.75	40.45
33.9									33.9	41.6
49.3										49.3

There are 55 averages, so the median is the 28th in order of increasing magnitude. Therefore, $\hat{\mu} = 29.5$.

37 Let $c = \frac{\Gamma((n-1)/2)}{\Gamma(n/2) \cdot \sqrt{2/(n-1)}}$. Then $E(cs) = cE(S)$, and c cancels with the two Γ factors and the square root in $E(S)$, leaving just σ. When n = 20, $c = \frac{\Gamma(9.5)}{\Gamma(10) \cdot \sqrt{2/19}}$. $\Gamma(10) = 9!$ and $\Gamma(9.5) = (8.5)(7.5)...(1.5)(.5)\,\Gamma(.5)$, but $\Gamma(.5) = \sqrt{\pi}$. Straightforward calculation gives c = 1.0132.

CHAPTER 7

Section 7.1

1 **a** $z_{\alpha/2} = 2.81$ implies that $\frac{\alpha}{2} = 1 - \Phi(2.81) = .0025$, so $\alpha = .005$ and the confidence level is $100(1-\alpha)\% = 99.5\%$.

b $1.44 = z_{\alpha/2}$ for $\alpha = 2[1 - \phi(1.44)] = .15$, and $100(1-\alpha)\% = 85\%$.

c 99.7% implies that $\alpha = .003$, $\frac{\alpha}{2} = .0015$, and $z_{.0015} = 2.96$ (look for cumulative area .9985 in the main body of Table A.3, the Z table).

d 75% implies $\alpha = .25$, $\frac{\alpha}{2} = .125$, and thus $z_{\alpha/2} = z_{.125} = 1.15$.

3 **a** $4.85 \pm \frac{(1.96)(.75)}{\sqrt{20}} = 4.85 \pm .33 = (4.52,\ 5.18)$

b $z_{\alpha/2} = z_{.02/2} = z_{.01} = 2.33$, so the interval is $4.56 \pm \frac{(2.33)(.75)}{\sqrt{16}} = (4.12,\ 5.00)$.

c $n = \left[\frac{2(1.96)(.75)}{.40}\right]^2 = 54.02$, so $n = 55$.

5 If $L = 2z_{\alpha/2}\frac{\sigma}{\sqrt{n}}$ and we increase the sample size by a factor of 4, the new length is $L' = 2z_{\alpha/2}\frac{\sigma}{\sqrt{4n}} = \left[2z_{\alpha/2}\frac{\sigma}{\sqrt{n}}\right]/2 = \frac{L}{2}$. Thus halving the length requires n to be increased fourfold. If $n' = 25n$, then $L' = \frac{L}{5}$, so the length is decreased by a factor of 5.

7 **a** $\left(\bar{x} - 1.645\frac{\sigma}{\sqrt{n}},\ \infty\right)$

from 3a, $\bar{x} = 4.85$, $\sigma = 0.75$, $n = 20$

$4.85 - (1.645)\frac{.75}{\sqrt{20}} = 4.5741$

interval: $(4.57,\ \infty)$

b $\left(\bar{x} - z_{\alpha}\frac{\sigma}{\sqrt{n}},\ \infty\right)$

Chapter 7

c $\left(-\infty,\ \bar{x}+z_{\alpha}\dfrac{\sigma}{\sqrt{n}}\right)$

from 2a, $\bar{x} = 58.3$, $\sigma = 3.0$, $n = 25$

$$58.3+2.33\left(\frac{3.0}{\sqrt{25}}\right) = (-\infty,\ 59.70)$$

9 Y is a binomial r.v. with $n = 1000$ and $p = .95$, so $E(Y) = np = 950$, the expected number of intervals that capture μ, and $\sigma_Y = \sqrt{npq} = 6.892$. Using the normal approximation to the binomial distribution,

$P(940 \le Y \le 960) = P(939.5 \le \text{normal variable} \le 960.5)$

$= P(-1.52 \le Z \le 1.52) = \Phi(1.52) - \Phi(-1.52) = .8714.$

Section 7.2

11 $n = 56$, $\bar{x} = 8.17$, $s = 1.42$

95% C.I. $\quad z_{\alpha/2} = 1.96$

$$8.17 \pm (1.96)\left(\frac{1.42}{\sqrt{56}}\right)$$

(7.798, 8.542)

We make no assumptions about the distribution of percentage elongation.

13 $$152.3 \pm \frac{(2.58)(4.8)}{\sqrt{35}} = 152.3 \pm 2.1 = (150.2,\ 154.4)$$

15 $\hat{p} = \dfrac{69}{277} = .249$, so the interval is $.249 \pm 2.58\left[\dfrac{(.249)(.751)}{277}\right]^{1/2}$

$= .249 \pm .067 = (.182,\ .316)$.

17 **a** Using (7.12) with $\hat{p} = .5$ yields $n = 385$.

b Again using (7.12) with $\hat{p} = \dfrac{2}{3}$ yields $n = 342$.

Chapter 7

19 With $\theta = \lambda$, $\hat{\theta} = \overline{X}$ and $\sigma_{\hat{\theta}} = \sqrt{\dfrac{\lambda}{n}}$ so $\hat{\sigma}_{\hat{\theta}} = \sqrt{\dfrac{\overline{X}}{n}}$. The large sample CI is then

$\overline{x} \pm z_{\alpha/2}\sqrt{\dfrac{\overline{x}}{n}}$. We calculate $\sum x_i = 203$ so $\overline{x} = 4.06$ and a 95% interval for λ is

$$4.06 \pm (1.96)\sqrt{\frac{4.06}{50}} = 4.06 \pm .56 = (3.50,\ 4.62).$$

Section 7.3

21

a 1.321 captures upper tail area .10, so cumulative area .90, and is thus the 90th percentile.

b Because any t curve is symmetric about zero,
10th percentile = −90th percentile = −1.321.

c 1.746 captures upper tail area .05, so −1.746 captures lower tail area .05.

d Area .01 lies to the right of 3.365, so
$P(T \le 3.365) = 1 - P(T > 3.365) = 1 - .01 = .99.$

e $P(-2.306 < T < 2.306) = 1 - [P(T \le -2.306) + P(T \ge 2.306)]$
$= 1 - [.025 + .025] = .95$

f $P(|T| \ge 3.435) = P(T \ge 3.435) + P(T \le -3.435) = .001 + .001 = .002$

23

a $\overline{x} + 9.52,\ s = 1.0686,\ n = 5$
for a 95% C.I. $t_{0.025,4} = 2.776$

$$9.52 \pm (2.776)\frac{1.0686}{\sqrt{5}} = (8.19,\ 10.85)$$

b $\left(-\infty,\ \overline{x} + (t_{\alpha,n-1})\dfrac{s}{\sqrt{n}}\right)$ $\quad t_{0.05,4} = 2.132$

$$\left(9.52 + 2.132\frac{1.0686}{\sqrt{5}}\right) = 10.54$$

$(-\infty, 10.54)$

25 $n = 9,\ \overline{x} = 188.0,\ s = 7.2$
$t_{0.025,8} = 2.306$

$$\overline{x} \pm t_{\alpha/2,n-1} \cdot s\sqrt{1 + \frac{1}{n}}$$

$$188.0 \pm (2.306)(7.2)\sqrt{1+\frac{1}{9}}$$

(170.5, 205.5)

27 $n = 18$, $\bar{x} = 38.66$, $s = 8.473$, $t_{.01,17} = 2.567$, so the 98% CI is $38.66 \pm 5.13 = (33.53, 43.79)$.

29 $\bar{x} = 21.90$, $s = 4.134$, $n = 10$

$$E(\bar{X}-\bar{X}^*) = E(\bar{X}) - E(\bar{X}^*) = \mu - \mu = 0$$

$$V(\bar{X}-\bar{X}^*) = V(\bar{X}) + V(\bar{X}^*) = \frac{\sigma^2}{n} + \frac{\sigma^2}{2} = \sigma^2\left(\frac{1}{n}+\frac{1}{2}\right)$$

$$\bar{x} \pm t_{\alpha/2} \cdot s\sqrt{\frac{1}{n}+\frac{1}{2}}$$

$$21.90 \pm (2.262)(4.134)\sqrt{\frac{1}{10}+\frac{1}{2}}$$

(14.66, 29.14)

Section 7.4

31

a $\chi^2_{.05,10} = 18.307$

b $\chi^2_{.95,10} = 3.940$

c Since $10.98 = \chi^2_{.975,22}$ and $36.78 = \chi^2_{.025,22}$, $P(\chi^2_{.975,22} \le \chi^2 \le \chi^2_{.025,22}) = .95$.

d Since $14.61 = \chi^2_{.95,\,25}$ and $37.65 = \chi^2_{.05,\,25}$, $P(\chi^2_{.95,\,25} \le \chi^2 \le \chi^2_{.05,\,25}) = .90$

33 $n = 22$ implies that d.f. $= n-1 = 21$, so the .995 and .005 columns of Table A.6 give the necessary chi-squared critical values as 8.033 and 41.399. $\sum x_i = 1701.3$ and $\sum x_i^2 = 132{,}097.35$, so $s^2 = 25.368$. The interval for σ^2 is $\left(\frac{21(25.368)}{41.399}, \frac{21(25.368)}{8.033}\right) = (12.868, 66.317)$, and that for σ is (3.6,8.1). Validity of this interval requires that fracture toughness be (at least approximately) normally distributed.

Chapter 7

Supplementary

35 We assume that histamine content has a normal distribution. With $\bar{x} = 638.1$, $s = 201.7$, $t_{.05,6} = 1.943$, the 90% CI is $638.1 \pm 148.1 = (490.0, 786.2)$.

37

a $\hat{p} = .680$ so the 90% CI for p is $.680 \pm 1.645\left[\dfrac{(.680)(.320)}{200}\right]^{1/2} = .680 \pm .054$ $= (.626, .734)$.

b Using $\hat{p} = \hat{q} = .5$, $L = \dfrac{2(1.645)}{\sqrt{4n}} = .05$, which yields $n = 1083$.

c 82% confidence is obtained by using $z_{.09} = 1.34$, whence the interval is $(.636, .724)$.

39 With $\hat{\theta} = \frac{1}{3}(\overline{X}_1 + \overline{X}_2 + \overline{X}_3) - \overline{X}_4$, $\sigma_{\hat{\theta}}^2 = \frac{1}{9}\text{Var}(\overline{X}_1 + \overline{X}_2 + \overline{X}_3) + \text{Var}(\overline{X}_4) = \frac{1}{9}\left[\frac{\sigma_1^2}{n_1} + \frac{\sigma_2^2}{n_2} + \frac{\sigma_3^2}{n_3}\right] + \frac{\sigma_4^2}{n_4}$; $\hat{\sigma}_{\hat{\theta}}$ is obtained by replacing each σ_i^2 by s_i^2 and taking the square root. The large-sample interval for θ is then $\frac{1}{3}(\bar{x}_1 + \bar{x}_2 + \bar{x}_3) - \bar{x}_4 \pm z_{\alpha/2}\left[\frac{1}{9}\left(\frac{s_1^2}{n_1} + \frac{s_2^2}{n_2} + \frac{s_3^2}{n_3}\right) + \frac{s_4^2}{n_4}\right]^{1/2}$. For the given data $\hat{\theta} = -.50$, $\hat{\sigma}_{\hat{\theta}} = .1718$, so the interval is $-.50 \pm (1.96)(.1718) = (-.84, -.16)$.

41 The specified condition is that the interval length be .2, so $n = \left[\dfrac{2(1.96)(.8)}{.2}\right]^2$ $= 245.86$; thus $n = 246$ should be used.

43 Proceeding as in Example 7.5 with T_r replacing $\sum X_i$, the CI for $\dfrac{1}{\lambda}$ is

$\left(\dfrac{2t_r}{\chi^2_{1-\alpha/2,2r}}, \dfrac{2t_r}{\chi^2_{\alpha/2,2r}}\right)$, where $t_r = y_1 + \dots + y_r + (n-r)y_r$. In Example 6.7, $n = 20$, $r = 10$, and $t_r = 1115$. With df = 20, the necessary critical values are 9.591 and 34.170, giving the interval (65.3, 232.5). This is obviously an extremely wide interval.

Chapter 7

45 **a** $\int_{(\alpha/2)^{1/n}}^{(1-\alpha/2)^{1/n}} nu^{n-1}du = u^n\Big]_{(\alpha/2)^{1/n}}^{(1-\alpha/2)^{1/n}} = 1 - \frac{\alpha}{2} - \frac{\alpha}{2} = 1 - \alpha$. From the probability statement, $\frac{(\alpha/2)^{1/n}}{\max(X_i)} \le \frac{1}{\theta} \le \frac{(1-\alpha/2)^{1/n}}{\max(X_i)}$ with probability $1 - \alpha$, so taking the reciprocal of each endpoint and interchanging gives the C.I. $\left(\frac{\max(x_i)}{(1-\alpha/2)^{1/n}}, \frac{\max(x_i)}{(\alpha/2)^{1/n}}\right)$ for θ.

b $\alpha^{1/n} \le \frac{\max(X_i)}{\theta} \le 1$ with probability $1 - \alpha$, so $1 \le \frac{\theta}{\max(X_i)} \le \frac{1}{\alpha^{1/n}}$ with probability $1 - \alpha$, which yields the interval $\left(\max(x_i), \frac{\max(x_i)}{\alpha^{1/n}}\right)$.

c It is easily verified that the interval of **b** is shorter — draw a graph of $f_U(u)$ and verify that the shortest interval that captures area $1 - \alpha$ under the curve is the rightmost such interval, which leads to the CI of **b**. With $\alpha = .05$, $n = 5$, $\max(x_i) = 4.2$; this yields (4.2, 7.65).

47 $\tilde{x} = 76.2$, the lower and upper fourths are 73.5 and 79.7, respectively, and $f_s = 6.2$. The robust interval is $76.2 \pm (1.93)\left(\frac{6.2}{\sqrt{22}}\right) = 76.2 \pm 2.6 = (73.6, 78.8)$.

$\bar{x} = 77.33$, $s = 5.037$, and $t_{.025,21} = 2.080$, so the t interval is $77.33 \pm \frac{(2.080)(5.037)}{\sqrt{22}} = 77.33 \pm 2.23 = (75.1, 79.6)$. The t interval is centered at $\bar{x}$, which is pulled out to the right of $\tilde{x}$ by the single mild outlier 93.7; the interval widths are comparable.

49 **a** $2126.5 \pm (2.4)(370.57) = 2126.5 \pm 889.4 = (1237.1, 3015.9)$

b If a large number of tolerance intervals are calculated from independent samples, approximately 95% of these will be such that at least 90% of all population values are included between the two interval limits.

c The width of the T.I. is $2(2.4)s$ whereas the width of the CI is $2t_{.025,16}\frac{s}{\sqrt{17}}$. The CI is much narrower than the T.I.

d k approaches 1.645 and the width approaches $2(1.645)\,\sigma$, whereas the width of the CI approaches zero.

CHAPTER 8

Section 8.1

1 **a** Yes. It is an assertion about the value of a parameter.

b No. The sample mean $\overline{x}$ is not a parameter.

c No. The sample standard deviation s is not a parameter.

d Yes. The assertion is that the standard deviation of population #2 exceeds that of population #1.

e No. $\overline{x}$ and $\overline{y}$ are statistics rather than parameters, so cannot appear in a hypothesis.

f Yes. H is an assertion about the value of a parameter.

3 In this formulation, H_0 states the welds do not conform to specification. This assertion will not be rejected unless there is strong evidence to the contrary. Thus the burden of proof is on those who wish to assert that the specification is satisfied. Using H_a: $\mu < 100$ results in the welds being believed in conformance unless provided otherwise, so the burden of proof is on the non-conformance claim.

5 Let σ denote the population standard deviation. The appropriate hypotheses are H_0: $\sigma = .05$ vs. H_a: $\sigma < .05$. With this formulation, the burden of proof is on the data to show that the requirement has been met (the sheaths will not be used unless H_0 can be rejected in favor of H_a).

Type I error: Conclude that the standard deviation is < .05 mm when it is really equal to .05 mm.

Type II error: Conclude that the standard deviation is .05 mm when it really is < .05 mm.

7 A type I error here involves saying that the plant is not in compliance when in fact it is. A type II error occurs when we conclude that the plant is in compliance when in fact it isn't. Reasonable people may disagree as to which of the two errors is more serious. If in your judgement it is the type II error, then the reformulation H_0: $\mu = 150$ vs. H_a: $\mu < 150$ makes the type I error more serious.

9 **a** H_0: $\mu = 1300$ vs. H_1: $\mu > 1300$

b $\overline{x}$ is normally distributed with mean $E(\overline{x}) = \mu$ and standard deviation $\frac{\sigma}{\sqrt{n}} = \frac{60}{\sqrt{20}} = 13.416$. When H_0 is true, $E(\overline{x}) = 1300$. Thus

$\alpha = P(\overline{x} \geq 1331.26$ when H_0 is true$) =$

$P(z \geq 1331.26 - 1300)/13.416) = P(z \geq 2.33) = .01$

c When $\mu = 1350$, $\bar{x}$ has a normal distribution with mean 1350 and standard deviation 13.416, so $\beta(1350) = P(\bar{x} < 1331.26 \text{ when } \mu = 1350) = P(z < (1331.26 - 1350)/13.416) = P(z < -1.40) = 0.0808$

d Replace 1331.26 by c, where c satisfies $\frac{c - 1300}{13.416} = 1.645$ (since $P(z \geq 1.645) = .05$). Thus $c = 1322.07$. Increasing α gives a decrease in β; now $\beta(1350) = P(z \leq -2.08) = .0188$.

e $\bar{x} \geq 1331.26$ iff $z \geq \frac{1331.26 - 1300}{13.416}$ i.e. iff $z \geq 2.33$.

11 a Let μ = true average braking distance for the new design at 40 mph. The hypotheses are H_0: $\mu = 120$ vs. H_a: $\mu < 120$.

b R_2 should be used, since support for H_a is provided only by an $\bar{x}$ value substantially smaller than 120 ($E(\bar{X}) = 120$ when H_0 is true and < 120 when H_a is true).

c $\sigma_{\bar{X}} = \frac{\sigma}{\sqrt{n}} = \frac{10}{6} = 1.6667$, so $\alpha = P(\bar{X} \leq 115.20 \text{ when } \mu = 120)$ $= P(z \leq (115.20 - 120)/1.6667) = P(z \leq -2.88) = 0.002$. To obtain $\alpha = .001$, replace 115.20 by $c = 120 - (3.08)(1.6667) = 114.87$, so that $P(\bar{X} \leq 114.87 \text{ when } \mu = 120) = P(z \leq -3.08) = .001$.

d $\beta(115) = P(\bar{X} > 115.20 \text{ when } \mu = 115) = P(z > .12) = .4522$

e $\alpha = P(z \leq -2.33) = 0.01$, because when H_0 is true Z has a standard normal distribution ($\bar{X}$ has been standardized using 120). Similarly, $P(Z \leq -2.88) = .002$, so this second rejection region is equivalent to R_2.

13 a $\sigma_{\bar{X}} = .04$, so $P(\bar{X} \geq 10.1004 \text{ or } \leq 9.8940 \text{ when } \mu = 10)$ $= P(z \geq 2.51 \text{ or } \leq -2.65) = .006 + .004 = .01$

b $\beta(10.1) = P(9.8940 < \bar{X} < 10.1004 \text{ when } \mu = 10.1)$ $= P(-5.15 < z < .01) = .5040$, whereas $\beta(9.9)$ $= P(-0.15 < z < 5.01) = 0.5596$. Since $\mu = 9.9$ and $\mu = 10.1$ represent equally serious departures from H_0, one would probably want to use a test procedure for which $\beta(9.9) = \beta(10.1)$. A similar result and comment apply to any other pair of alternative values symmetrically placed about 10.

Section 8.2

15 a $\alpha = P(t \geq 3.733$ when t has a t distribution with 15 df$) = .001$, because the 15 df row of Table A.5 shows that $t_{.001,15} = 3.733$.

b df = $n - 1 = 23$, so $\alpha = P(t \leq -2.500) = .01$

c df = 30, and $\alpha = P(t \geq 1.697) + P(t \leq -1.697) = .05 + .05 = .10$

Chapter 8

17 **a** $\frac{72.3-75}{1.8} = -1.5$, so 72.3 is 1.5 SD's (of $\bar{x}$) below 75.

b H_0 is rejected if zZ $\le$ -2.33; since $z = -2.87 \le -2.33$, reject H_0.

c α = area under standard normal curve below -2.88 = $\Phi(-2.88) = .0020$

d $\Phi\left(-2.88 + \frac{75-70}{9/5}\right) = \Phi(-.1) = .4602$, so $\beta(70) = .5398$

e $n = \left[\frac{9(2.88+2.33)}{75-70}\right]^2 = 87.95$, so use $n = 88$.

f $\alpha(76) = P(Z < -2.33 \text{ when } \mu = 76) = P(\bar{X} < 72.9 \text{ when } \mu = 76)$

$= \Phi\left(\frac{72.9-76}{.9}\right) = \Phi(-3.44) = .0003$

19 **a** H_0: $\mu = 250$ vs. H_a: $\mu > 250$, reject H_0 if zZ $\ge$ 1.645; since $\frac{\sigma}{\sqrt{n}} = \frac{15}{\sqrt{20}} =$ 3.35, $z = \frac{257.3-250}{3.35} = 2.18 \ge 1.645$, so reject H_0.

b $\Phi(1.645 + (10/3.35)) = \Phi(-1.34) = .0901$.

c Should test H_0: $\mu = 250$ (actually ≥ 250) vs. H_a: $\mu < 250$, since the burden of proof is then on H_a, the manufacturer's claim.

21 Reject H_0 if $z \ge 1.645$; $\frac{s}{\sqrt{n}} = .811$, so $z = \frac{52.7-50}{0.811} = 3.33$. Since 3.33 is $\ge$ 1.645, reject H_0 at level .05 and conclude that true average penetration exceeds 50 mils.

23 We wish to test H_0: $\mu = 75$ vs. H_a: $\mu < 75$. Using $\alpha = .01$, H_0 is rejected if $z \le -2.33$. Since $z = \frac{73.1-75}{5.9/\sqrt{32}} = -1.82$, which is not $\le$ -2.33, H_0 is not rejected. $P = \Phi(-1.82) = .0344$.

25 With μ = true average recumbency time, the hypotheses are H_0: $\mu = 20$ vs. H_a: $\mu < 20$. The test statistic value is $z = \frac{\bar{x}-20}{s/\sqrt{n}}$, and H_0 should be rejected if $z \le -z_{.10} = -1.28$. Since $z = \frac{18.86-20}{8.6/\sqrt{73}} = \frac{-1.14}{1.007} = -1.13$, which is not $\le$ -1.28, H_0 is not rejected. The sample data does not strongly suggest that true average time is less than 20.

Chapter 8

27 $n = 115,\ \bar{x} = 11.3,\ s = 6.43$

1. Parameter of interest: $\boldsymbol{\mu}$ = true average dietary intake of zinc among males aged 65 - 74 years.
2. Null hypothesis: H_0: $\mu = 15$
3. Alternative hypothesis: H_a: $\mu < 15$
4. $$z = \frac{\bar{x} - \mu_0}{s/\sqrt{n}} = \frac{\bar{x} - 15}{s/\sqrt{n}}$$
5. Rejection region: No value of $\boldsymbol{\alpha}$ was given, so select a reasonable level of significance, such as $\boldsymbol{\alpha} = .05$

   $z \le z_\alpha$ or $z \le -1.645$
6. $$z = \frac{11.3 - 15}{6.43/\sqrt{115}} = -6.17$$
7. $-6.17 < -1.645$ so reject H_0. The data does support the claim that average daily intake of zinc for males aged 65 - 74 years falls below the recommended daily allowance of 15 mg/day.

29 Let μ denote the true average soil pH after treatment. We wish to test H_0: $\mu = 8.75$ vs. H_a: $\mu \ne 8.75$. A two-tailed t test will be used, with

$$t = \frac{\bar{x} - 8.75}{s/\sqrt{n}};\ t_{.005,4} = 4.604,$$ so H_0 will be rejected if either $t \ge 4.604$ or ≤ -4.604.

Because $t = \dfrac{-.75}{.0224} = -33.5 < -4.604$, H_0 is rejected. The effluent treatment does appear to affect true average pH.

31 $$\beta(\mu_0 - \Delta) = \Phi(z_{\alpha/2} + \Delta\sqrt{n}/\sigma) - \Phi(-z_{\alpha/2} - \Delta\sqrt{n}/\sigma) = 1 - [(-z_{\alpha/2} - \Delta\sqrt{n}/\sigma) + \Phi(z_{\alpha/2} - \Delta\sqrt{n}/\sigma)]$$
(since $1 - \Phi(c) = \Phi(-c)$) $= \beta(\mu_0 + \Delta)$.

Section 8.3

33

1. Parameter of interest: p = true proportion of cars in this particular county passing emissions testing on the first try.
2. H_0: $p = 0.70$
3. H_a: $p \ne 0.70$
4. $$z = \frac{\hat{p} - p_0}{\sqrt{(p_0)(1 - p_0)/n}} = \frac{\hat{p} - .70}{\sqrt{(.30)(.70)/n}}$$
5. either $z \ge 1.96$ or $z \le -1.96$

Chapter 8

6 $z = \dfrac{156/200 - .70}{\sqrt{(.70)(.30)/200}} = 2.469$

7 Reject H_0, the data indicates that the proportion of cars passing the first time on emissions testing for this county differs from the proportion of cars passing statewide.

35 H_0: $p = .6$ vs. H_a: $p > .6$ are the appropriate hypotheses, so H_0 is rejected (and the installation carried out) only if $z \geq 1.645$. With $\hat{p} = .7375$, $z = \dfrac{.7375 - .6}{.0387} = 3.55$. Since $3.55 \geq 1.645$, reject H_0 at level .05.

37 Let p denote the true proportion of those called to appear for service who are black. H_0: $p = .25$ vs. H_a: $p < .25$ will be tested using $\dfrac{\hat{p} - .25}{((.25)(.75)/n)^{.5}}$ and rejection region $z \leq -z_{.01} = -2.33$. We calculate $\hat{p} = \dfrac{177}{1050} = .1686$ and $z = \dfrac{-.0814}{.0134} = -6.1$ Because $-6.1 \leq -2.33$, H_0 is rejected. A conclusion that discrimination exists is very compelling.

39 The hypotheses are H_0: $p = .10$ vs. H_a: $p > .10$ so R has the form $\{c,\ldots,n\}$. For $n = 10$, $c = 3$ (i.e. $R = \{3, 4,\ldots, 10\}$) yields $\alpha = 1 - B(2;\ 10,\ .1) = .07$ while no larger R has $\alpha \leq .10$; however, $\beta(.3) = B(2;\ 10,\ .3) = .383$. For $n = 20$, $c = 5$ yields $\alpha = 1 - B(4;\ 20,\ .1) = .043$, but again $\beta(.3) = B(4;\ 20,\ .3) = .238$. For $n = 25$, $c = 5$ yields $= 1 - B(4;\ 25,\ .1) = .098$ while $\beta(.7) = B(4;\ 25,\ .3) = .090 \leq .10$, so $n = 25$ should be used.

41 H_0: $p = .03$ vs. H_a: $p < .03$, reject H_0 if $z \leq -2.33$ where $z = \dfrac{\hat{p} - .03}{\sqrt{(.03)(.97)/n}}$, with $\hat{p} = .028$, $z = \dfrac{-.002}{.00763} = -.26$; $-.26$ isn't ≤ -2.33, so H_0 is not rejected. Robots have not demonstrated their superiority.

Section 8.4

43

- **a** P-value $= .084 > .05 = \alpha$, so don't reject H_0
- **b** P-value $= .003 > .001 = \alpha$, so don't reject H_0
- **c** $.498 >> .05$ so H_0 cannot be rejected at significance level .05
- **d** $.084 < .10$, so reject H_0 at level .10
- **e** .039 is not $< .01$, so don't reject H_0
- **f** P-value $= .218 > .10 = \alpha$, so H_0 cannot be rejected.

Chapter 8

45 **a** .0358
b .0802
c .5824
d .1586
e 0

47 **a** $t_{.05,6} = 1.943 < 2.3 < 2.447 = t_{.025,6}$, but for a two-tailed test the subscripts refer to $\alpha/2$, so 2(.025) < P-value < 2(.05), i.e., .05 < P-value < .10
b $-t_{.005,14} = -2.977 \approx -3$, so P-value $\approx$ 2(.005) = .01
c $t_{.0005,24} = 3.745 < 4.2$, so P-value < 2(.0005) = .001
d $-1.3 \approx -1.333 = -t_{.10,17}$, so P-value $\approx$ 2(.10) = .20

49 Here we might be concerned with departures above as well as below the specified weight of 5.0, so the relevant hypotheses are H_0: $\mu = 5.0$ vs. H_a: $\mu \neq 5.0$. At level .01, reject H_0 if either $z \geq 2.58$ or if $z \leq -2.58$. Since $\frac{s}{\sqrt{n}} = .035$, $z = \frac{-.13}{.035} = -3.71 \leq$ - 2.58, so H_0 can be rejected. Because 3.71 is "off" the z table, P < 2(.0002) = .0004 (.0002 corresponds to z = -3.49).

51 The hypotheses to be tested are H_0: $\mu = 25$ vs. H_a: $\mu > 25$, and H_0 should be rejected if $t \geq t_{.05,12} = 1.782$. The computed summary statistics are $\bar{x} = 27.923$, s = 5.619, so $\frac{s}{\sqrt{n}} = 1.559$ and $t = \frac{2.923}{1.559} = 1.88$. Because $t_{.025,12} = 2.179 > 1.88 > t_{.05,12}$ = 1.782, the P-value is between .025 and .05, so H_0 is rejected at level .05. It is normal.

53 H_0: $\mu = 15$ vs. H_a: $\mu > 15$, $t = \frac{52.5 - 15}{8/\sqrt{11}} = 15.5 >> 4.587 = t_{.0005,10}$, so P-value << .0005. Thus H_0 is easily rejected at significance level .01. The average level does appear to exceed the standard.

Section 8.5

55 **a** Here $\beta = \Phi\left(\frac{-.01 + 9320/\sqrt{n}}{.4073/\sqrt{n}}\right) = \Phi\left(\frac{-.01\sqrt{n} + .9320}{.4073}\right)$ = .9703, .8554, .4325, .0944, and 1 for n = 100, 2500, 10,000, 40,000, and 90,000, respectively.
b Here $z = .025\sqrt{n}$ = .25, 1.25, 2.5, and 5 for the four n's, whence P = .4013, .1056, .0062, .0000, respectively.
c No; the reasoning is the same as in 54(c).

Chapter 8

Supplementary Exercises

57 Here we assume that thickness is normally distributed, so that for any n a t test is appropriate, and use Table A.13 to determine n. We wish $\pi(3) = .95$ when $d = \frac{|3.2-3|}{.3} = .667$. By inspection, $n = 20$ satisfies this requirement, so $n = 50$ is too large.

59

a The relevant hypotheses are H_0: $\mu = 548$ vs. H_a: $\mu \neq 548$. At level .05, H_0 will be rejected if either $t \geq t_{.025,10} = 2.228$ or $t \leq -2.228$. The test statistic value is $t = \frac{587-548}{10/\sqrt{11}} = \frac{39}{3.02} = 12.9$. This clearly falls in the upper tail of the two-tailed rejection region, so H_0 should be rejected at level .05 (or any other reasonable level).

b The population sampled was normal or approximately normal.

61 $n = 47$, $\bar{x} = 215$ mg, $s = 235$ mg
range: 5 mg to 1,176 mg

a No, the distribution does not appear to be normal, it appears to be skewed to the right. It is not necessary to assume normality if the sample size is large enough due to the central limit theorem. This sample size is large enough so we can conduct a hypothesis test about the mean.

b

1. Parameter of interest: μ = true daily caffeine consumption of adult women.
2. H_0: $\mu \leq 200$
3. H_a: $\mu > 200$
4. $z = \frac{\bar{x}-\mu_0}{s/\sqrt{n}} = \frac{\bar{x}-200}{s/\sqrt{n}}$
5. *RR*: $z \geq 1.282$ or if *P*-value $\leq .10$
6. $z = \frac{215-200}{235/\sqrt{47}} = 0.44$ *P*-value $= 1 - \Phi(.44) = .33$
7. Fail to reject H_0 because .33 > 0.10. The data does not indicate that daily caffeine consumption of adult women exceeds 200 mg.

63

a From Table A.13
when $\mu = 9.5$, $d = 0.625$, $df = 9$, $\beta \approx .60$
when $\mu = 9.0$, $d = 1.25$, $df = 9$, $\beta \approx .20$

b From Table A.13
$\beta = .25$, $d = .625$, $n = 28$

Chapter 8

65 Assuming that the weight gain distribution is normal, we use a t test, rejecting H_0: $\mu = 2.5$ in favor of H_a: $\mu > 2.5$ if $t \geq t_{\alpha,n-1} = t_{.001,15} = 3.733$. The computed T is $t = \dfrac{2.79 - 2.5}{.41/\sqrt{16}} = 2.83$. Because 2.83 is not ≥ 3.733, H_0 cannot be rejected at level .001.

67 **a** Assuming normality, a t test is appropriate; H_0: $\mu = 1.75$ is rejected in favor of H_a: $\mu \neq 1.75$ if either $t \geq t_{.025,25} = 2.060$ or $t \leq -2.060$. The computed t is $t = \dfrac{\sqrt{26}(1.89 - 1.75)}{.42} = 1.70$. Since 1.70 is neither ≥ 2.060 nor ≤ -2.060, do not reject H_0; the data does not contradict prior research.

b $1.70 \doteq 1.708 = t_{.05,25}$, so $P \doteq 2(.05) = .10$ (since for a two-tailed test, $.05 = \alpha/2$.

69 Let p = the true proportion of mechanics who could identify the problem. Then the appropriate hypotheses are H_0: $p = .75$ vs. H_a: $p < .75$, so a lower-tailed test should be used. With $p_0 = .75$ and $\hat{p} = \dfrac{42}{72} = .583$, $z = -3.28$ and $P = \Phi(-3.27) = .0005$. Because this P-value is very small, the data argues strongly against H_0, so we reject it in favor of H_a.

71 H_0: $\mu = 15$ vs. H_a: $\mu > 15$. $z = \dfrac{\bar{x} - 15}{s/\sqrt{n}} = \dfrac{17.5 - 15}{2.2/\sqrt{32}} = \dfrac{2.5}{.390} = 6.4$. Thus P-value $= 1 - \Phi(6.4) = 0 < .05$, so H_0 is rejected in favor of the conclusion that the true average time exceeds 15 min.

73 The 20 df row of Table A.6 shows that $\chi^2_{.99,20} = 8.26 < 8.58$ (H_0 not rejected at level .01) and $8.58 < 9.591 = \chi^2_{.975,20}$ (H_0 rejected at level .025). Thus $.01 < P\text{-value} < .025$ and H_0 cannot be rejected at level .01 (the P-value is the smallest α at which rejection can take place, and this exceeds .01).

75 **a** When H_0 is true, $2\lambda_0 \Sigma X_i = 2\Sigma \dfrac{X_i}{\mu_0}$ has a chi-squared distribution with df = $2n$. If the alternative is H_a: $\mu > \mu_0$, large test statistic values (large Σx_i, since $\bar{x}$ is large) suggest that H_0 be rejected in favor of H_a, so rejecting when $\dfrac{2\Sigma x_i}{\mu_0} \geq \chi^2_{\alpha,2n}$ gives a test with significance level α. If the alternative is

H_a: $\mu < \mu_0$, rejecting when $2\frac{\Sigma x_i}{\mu_0} \le \chi^2_{1-\alpha,2n}$ gives a level α test. The rejection region for H_a: $\mu \ne \mu_0$ is either $2\frac{\Sigma x_i}{\mu_0} \ge \chi^2_{\alpha/2,2n}$ or $\le \chi^2_{1-\alpha/2,2n}$.

b H_0: $\mu = 75$ vs. H_a: $\mu < 75$. The test statistic value is $\frac{2(737)}{75} = 19.65$. At level .01, H_0 is rejected if $2\frac{\Sigma x_i}{\mu_0} \le \chi^2_{.99,20} = 8.260$. Clearly 19.65 is not in the rejection region, so H_0 should not be rejected. The sample data does not suggest that true average lifetime is less than the previously claimed value.

77 **a** $\alpha = P(X \le 5$ when $p = .9) = B(5;\ 10,\ .9) = .002$, so the region (0,1,...5) does specify a level .01 test.

b The first value to be placed in the upper-tailed part of a two-tailed region would be 10, but $P(X = 10$ when $p = .9) = .349$, so whenever 10 is in the rejection region, $\alpha \ge .349$.

CHAPTER 9

Section 9.1

1 **a** $E(\overline{X}-\overline{Y}) = E(\overline{X}) - E(\overline{Y}) = 4.1 - 4.5 = -.4$, irrespective of the sample sizes.

b $V(\overline{X}-\overline{Y}) = V(\overline{X}) + V(\overline{Y}) = \dfrac{\sigma_1^2}{m} + \dfrac{\sigma_2^2}{n} = \dfrac{(1.8)^2}{100} + \dfrac{(2.0)^2}{100} = .0724$, and the sd of $\overline{X}-\overline{Y}$ is $\sqrt{.0724} = .2691$.

c A normal curve with mean and sd as given in **a** and **b** (because $m = n = 100$, the CLT implies that both $\overline{X}$ and $\overline{Y}$ have approximately normal distributions, so $\overline{X}-\overline{Y}$ does also). The shape is not necessarily that of a normal curve when $m = n = 10$, because the CLT cannot be invoked. So if the two lifetime population distributions are not normal, the distribution of $\overline{X}-\overline{Y}$ will typically be quite complicated.

3 The test statistic value is $z = \dfrac{(\bar{x}-\bar{y}-10{,}000)}{(s_1^2/m + s_2^2/n)^{1/2}}$, and H_0 will be rejected at level .01 if $z \ge 2.33$. We compute $z = \dfrac{2700}{(177{,}250)^{1/2}} = \dfrac{2700}{421.01} = 6.41$. Because $6.41 \ge 2.33$, reject H_0 and conclude that true average life for radials exceeds that for economy brand by more than 10,000.

5 **a** $\left(\dfrac{\sigma_1^2}{m} + \dfrac{\sigma_2^2}{n}\right)^{1/2} = \left(\dfrac{.04}{10} + \dfrac{.16}{10}\right)^{1/2} = .1414$, so $z = \dfrac{(.64 - 2.05 - (-1))}{.1414} = -2.90$. At level .01, H_0 is rejected if $z \le -2.33$; since $-2.90 \le -2.33$, reject H_0.

b $P = \Phi(-2.90) = .0019$

c $\beta = 1 - \Phi\left(-2.33 - \dfrac{-1.2+1}{.1414}\right) = 1 - \Phi(-.92) = .8212$

d $m = n = \dfrac{.2(2.33+1.28)^2}{(-.2)^2} = 65.16$, so use 66.

7 **1** Parameter of interest: $\mu_1 - \mu_2$ = the true difference of means for males and females on the Boredom Proneness Rating. Let μ_1 = men's average and μ_2 = women's average.

2 H_0: $\mu_1 - \mu_2 = 0$

3 H_a: $\mu_1 - \mu_2 > 0$

Chapter 9

4 $$z = \frac{(\bar{x}-\bar{y}) - \Delta_0}{\sqrt{\frac{s_1^2}{m}+\frac{s_2^2}{n}}} = \frac{(\bar{x}-\bar{y}) - 0}{\sqrt{\frac{s_1^2}{m}+\frac{s_2^2}{n}}}$$

5 RR: $z \geq 1.645$

6 $$z = \frac{10.40 - 9.26}{\sqrt{\frac{(4.83)^2}{97}+\frac{(4.68)^2}{148}}} = 1.83$$

7 Reject H_0. The data indicates the Boredom Proneness Rating is higher for males than for females.

9

a The hypotheses are H_0: $\mu_1 - \mu_2 = 5$ versus H_a: $\mu_1 - \mu_2 > 5$. At level .001, H_0 should be rejected if $z \geq 3.08$. Since $z = \frac{(65.6 - 59.8 - 5)}{.2772} = 2.89 < 3.08$, H_0 cannot be rejected in favor of H_a at this level, so the use of the high-purity steel cannot be justified.

b $\mu_1 - \mu_2 - \Delta_0 = 1$, so $\beta = \Phi\left(3.08 - \frac{1}{.2772}\right) = \Phi(-.53) = .2981$

c The computed Z and conclusion are identical to those of **a**.

11 The C.I. is $\bar{x} - \bar{y} \pm (2.58)\left(\frac{s_1^2}{m}+\frac{s_2^2}{n}\right)^{1/2} = -8.77 \pm (2.58)(.9104)^{1/2} = -8.77 \pm 2.46$ $= (-11.23, -6.31)$.

13 The appropriate hypotheses are H_0: $\theta = 0$ versus H_a: $\theta < 0$, where $\theta = 2\mu_1 - \mu_2$ ($\theta < 0$ is equivalent to $2\mu_1 < \mu_2$, so normal is more than twice schizo). The estimator of θ is $\hat{\theta} = 2\bar{X} - \bar{Y}$, with $\text{Var}(\hat{\theta}) = 4\ \text{Var}(\bar{X}) + \text{Var}(\bar{Y}) = \left(\frac{4\sigma_1^2}{m}+\frac{\sigma_2^2}{n}\right)$, $\sigma_{\hat{\theta}}$ the square root of $\text{Var}(\hat{\theta})$, and $\hat{\sigma}_{\hat{\theta}}$ obtained by replacing each σ_i^2 by S_i^2. The test statistic is then $\frac{\hat{\theta}}{\sigma_{\hat{\theta}}}$ (since $\theta_0 = 0$), and H_0 is rejected if $z \leq -2.33$. With $\hat{\theta} = 2(2.69) - 6.35 = -.97$ and $\hat{\sigma}_{\hat{\theta}} = \left(\frac{4(2.3)^2}{33}+\frac{(4.03)^2}{35}\right)^{1/2} = 1.05$, $z = \frac{-.97}{1.05} = -.92$; because $-.92 > -2.33$, H_0 is not rejected.

15 As β decreases, z_β increases, and since z_β is the numerator of n, n increases also.

Chapter 9

Section 9.2

17 The test statistic value is $t = \dfrac{\bar{x}-\bar{y}}{s_p(1/m+1/n)^{1/2}}$. The number of d.f. for the test is $m+n-2 = 6+5-2 = 9$, and $t_{.025,9} = 2.262$. H_0 will be rejected if either $t \geq 2.262$ or $t \leq -2.262$. The numerator of s_p^2 is $5(5.59)^2+4(5.25)^2 = 266.49$, so $s_p^2 = \dfrac{266.49}{9} = 29.61$ and $s_p = 5.44$. Thus $t = \dfrac{-6.6}{5.44(.1667+.2000)^{1/2}} = \dfrac{-6.6}{3.294}$ $= -2.00$. Because -2.00 is neither ≤ -2.262 nor ≥ 2.262, don't reject H_0. The data does not indicate that the two true average stopping distances are different.

19 $\bar{x}-\bar{y} \pm t_{\alpha/2,m+m-2} \cdot s_p\sqrt{\dfrac{1}{m}+\dfrac{1}{n}}$

$(115.7-129.3) \pm (2.228)(5.2079)\sqrt{\dfrac{1}{6}+\dfrac{1}{6}}$

$(-20.3, -6.9)$

The interval is fairly wide, indicating a lack of precision.

21

a $s_p = \sqrt{\dfrac{5(11.3)^2+7(8.3)^2}{6+8-2}} = 9.66$

b The P-value = .0018 so reject H_0 at the level $\alpha = .01$.

c $t_{.025,12} = 2.179$ $\quad t_{.005,12} = 3.055$

The 95% C.I. is centered at $\dfrac{7.5+30.3}{2} = 18.9$

95% C.I.: $18.9 \pm t_{.025,12}$ (standard error)

$18.9 \pm (2.1749)$ (standard error)

Solving for standard error yields 5.2318

$18.9 \pm (3.055)(5.2318)$

(2.9, 34.9)

23

a H_0: $\mu_1-\mu_2 = 0$ is rejected in favor of H_a: $\mu_1-\mu_2 < 0$ if $t \leq -t_{.05,18}$ $= -1.734$. With $s_p^2 = \dfrac{(s_1^2+s_2^2)}{2} = 21{,}451.88$, $s_p = 146.46$ and $t = \dfrac{-160.8}{65.50} = -2.45 \leq -1.734$, so H_0 is rejected; the second type does appear to have greater true average bonding strength.

b Because $t_{.025,18} < 2.45 < t_{.01,18}$, $.01 < P < .025$.

c $d = \frac{|100-0|}{150} \bullet \sqrt{\frac{(10)(10)}{(20)(19)}} = .34$ and $m + n - 2 = 18$, so from Table A.13, $\beta \approx .54$.

d It appears as though $n = 12$ will suffice.

25 $m = 5$, $\bar{x} = 26.58$, $s_1 = 2.43$, $n = 5$, $\bar{y} = 40.24$, $s_2 = 2.93$. With $t_{.025,8} = 2.306$ and $s_p = 2.69$, the 95% C.I. for $\mu_1 - \mu_2$ is $-13.66 \pm 3.92 = (-17.58, -9.74)$.

27 Let μ_1 and μ_2 denote true average soil pH's at the two locations. We wish to test H_0: $\mu_1 - \mu_2 = 0$ versus H_a: $\mu_1 - \mu_2 \neq 0$. Assuming that the pH distributions are both normal with $\sigma_1 = \sigma_2$, the pooled t test based on $m+n-2 = 16$ d.f. will be used: reject H_0 if either $t \geq 2.120$ or $t \leq -2.120$. We calculate $\bar{x} = 8.038$, $s_1 = .2852$, $\bar{y} = 7.442$, $s_2 = .2244$, $s_p = .2567$, $t = \frac{.596}{(.2567)(.4714)} = \frac{.596}{.1210} = 4.93$. Because $4.93 \geq 2.120$, reject H_0. The two average pH's do appear to differ.

29 With the subscript 1 identified with the nitrate treatment, we wish to test H_0: $\mu_1 - \mu_2 = 0$ versus H_a: $\mu_1 - \mu_2 < 0$. The summary quantities are $m = 9$, $\bar{x} = 15.07$, $s_1 = 3.56$, $n = 7$, $\bar{y} = 19.27$, $s_2 = 8.05$. Thus

$$\nu = \frac{\left[\frac{12.67}{9} + \frac{64.80}{7}\right]^2}{\frac{(12.67/9)^2}{8} + \frac{(64.80/7)^2}{6}} = \frac{113.74}{14.53} = 7.83$$

so $\nu = 8$ and critical value $-t_{.01,8} = -2.896$. The computed value of T' is $t' = \frac{15.07 - 19.27}{\left(\frac{12.67}{9} + \frac{64.80}{7}\right)^{1/2}} = -1.29$. Since -1.29 is not ≤ -2.896, don't reject H_0.

Section 9.3

31 $\bar{d} = 7.25$ $\quad s_D = 11.8628$

1. Parameter of interest: μ_D = true average difference of breaking load for fabric in unabraded or abraded condition.
2. H_0: $\mu_D = 0$
3. H_a: $\mu_D = 0$

Chapter 9

4 $t = \dfrac{\bar{d} - \mu_D}{s_D/\sqrt{n}} = \dfrac{\bar{d} - 0}{s_D/\sqrt{n}}$

5 *RR*: $t_{.01,\,7}$ or $t > 2.998$

6 $t = \dfrac{7.25 - 0}{\dfrac{11.8628}{\sqrt{8}}} = 1.73$

7 Fail to reject H_0. The data does not indicate a difference in breaking load for the two fabric conditions.

33 H_0: $\mu_D = 0$ is rejected at level .1 in favor of H_a: $\mu_D \neq 0$ if either $t \geq t_{.05,13} = .1771$ or $t \leq -1.771$. The summary quantities are $\bar{d} = 1.21$, $s_D = 12.68$, so $t = \dfrac{1.21}{3.39} = .36$. Because $.36 < 1.771$, H_0 cannot be rejected.

35 The data is paired, so to obtain a 99% CI for μ_D we compute $\bar{d} = -.414$ and $s_D = .321$. With $t_{.005,7} = 3.499$, the interval is $-.414 \pm \dfrac{(3.499)(.321)}{\sqrt{8}} = -.414 \pm .397$ $= (-.811, -.017)$.

37 The data is paired and we wish a 95% CI for μ_D. The d_i's are 111, −65, 37, 69, 186, 59, 155, −10, and 253, whence $\bar{d} = 88.33$ and $s_D = 99.16$. With $t_{.025,8} = 2.306$, the 95% CI is $88.33 \pm 76.22 = (12.11, 164.55)$.

39 The differences (white – black) are −7.62, −8.00, −9.09, −6.06, −1.39, −16.07, −8.40, −8.89, and −2.88, from which $\bar{d} = -7.600$ and $s_D = 4.178$. The confidence level is not specified in the problem description; for 95% confidence, $t_{.025,8} = 2.306$ and the CI is $-7.600 \pm \dfrac{(2.306)(4.178)}{\sqrt{9}} = -7.600 \pm 3.211$ $= (-10.81, -4.39)$.

Section 9.4

41 H_0 will be rejected if $z \leq -z_{.01} = -2.33$. With $\hat{p}_1 = .150$, $\hat{p}_2 = .300$, $\hat{p} = \dfrac{30 + 180}{200 + 600}$ $= \dfrac{210}{800} = .263$, and $\hat{q} = .737$, the numerator of z is $.150 - .300 = -.150$, the

denominator is $\left[(.263)(.737)\left(\frac{1}{200}+\frac{1}{600}\right)\right]^{1/2} = .0359$, and $z = \frac{-.150}{.0359} = -4.18$.
Because $-4.18 \le -2.33$, H_0 is rejected; the proportion of those who repeat after inducement appears lower than those who repeat after no inducement.

43 **1** Parameter of interest: $p_1 - p_2$ = true difference in proportions of those responding to two different survey covers. Let p_1 = Plain, p_2 = Picture.

2 H_0: $p_1 - p_2 = 0$

3 H_a: $p_1 - p_2 < 0$

4
$$z = \frac{\hat{p}_1 - \hat{p}_2}{\sqrt{\hat{p}\hat{q}\left(\frac{1}{m}+\frac{1}{n}\right)}}$$

5 Reject H_o if P-value < 0.10

6
$$z = \frac{\frac{104}{207}-\frac{109}{213}}{\sqrt{\left(\frac{213}{420}\right)\left(\frac{207}{420}\right)\left(\frac{1}{207}+\frac{1}{213}\right)}} = -0.1910$$
P-value = 0.4247

7 Fail to reject H_0. The data does not indicate that plain cover surveys have a lower response rate.

45 **a** H_o: $p_1 = p_2$ will be rejected in favor of H_a: $p_1 \ne p_2$ if either $z \ge 1.645$ or $z \le -1.645$. With $\hat{p}_1 = .193$, $\hat{p}_2 = .182$, $\hat{p} = .188$, $z = \frac{.011}{.00742} = 1.48$. Since 1.48 is not $\ge$ 1.645, H_0 is not rejected and we conclude that no difference exists.

b Using formula (9.10) with $p_1 = .2$, $p_2 = .18$, $\alpha = .1$, $\beta = .1$ and $z_{\alpha/2} = 1.645$,
$$n = \frac{(1.645\sqrt{.5(.38)(1.62)}+1.28\sqrt{.16+.1476})^2}{.0004} = 6582.$$

47 $\hat{\theta} = \frac{\hat{p}_1}{\hat{p}_2}$ $\qquad V(\hat{\theta}) = \frac{p_1(1-p_1)}{n_1 p_2^2} + \frac{p_1^2(1-p_2)}{n_2 p_2^3}$

control = $\hat{p}_1 = \frac{189}{11{,}034}$, aspirin = $\hat{p}_2 = \frac{104}{11{,}037}$

$$\frac{\frac{189}{11,034}}{\frac{104}{11,037}} + 1.96\sqrt{\frac{\left(\frac{189}{11,034}\right)\left(\frac{10,845}{11,034}\right)}{(11,034)\left(\frac{104}{11,037}\right)^2} + \frac{\left(\frac{189}{11,034}\right)\left(\frac{10,933}{11,037}\right)}{(11,037)\left(\frac{104}{11,037}\right)^3}}$$

(1.385, 2.250)

The interval suggests the aspirin is more effective at reducing heart attacks.

49 $\hat{p}_1 = \frac{15+7}{40} = .550$, $\hat{p}_2 = \frac{29}{42} = .690$, and the 95% C.I. is $.550 - .690 \pm (1.96)(.106)$ $= -.14 \pm .21 = (-.35, .07)$.

Section 9.5

51

a From Table A.7, column 5, row 8, $F_{.05,5,8} = 3.69$

b From column 8, row 5, $F_{.05,8,5} = 4.82$

c $F_{.95,5,8} = \frac{1}{F_{.05,5,8}} = .207$

d $F_{.95,8,5} = \frac{1}{F_{.05,8,5}} = .271$

e $F_{.01,10,12} = 4.30$

f $F_{.99,10,12} = \frac{1}{F_{.01,12,10}} = \frac{1}{4.71} = .212$

g $F_{.05,6,4} = 6.16$, so $P(F \le 6.16) = .95$

h Since $F_{.99,10,5} = \frac{1}{5.64} = .177$, $P(.177 \le F \le 4.74)$ $= P(F \le 4.74) - P(F \le .177) = .95 - .01 = .94$.

53 H_0: $\sigma_1 = \sigma_2$ will be rejected at level .02 if either $f \ge F_{.01,11,11} \approx 4.47$ or $f \le \frac{1}{4.47} = .224$. Since $f = \left(\frac{.13}{.17}\right)^2 = .585$, which is neither ≥ 4.47 nor $\le .224$, H_0 is not rejected.

55 H_0: $\sigma_1 = \sigma_2$ will be rejected in favor of H_a: $\sigma_1 \ne \sigma_2$ if $f \le F_{.975,47,44} \approx .56$ or if $f \ge F_{.025,47,44} \approx 1.8$. Because $f = 1.22$, H_0 is not rejected.

Chapter 9

Supplementary

57
$$s_p = \sqrt{\frac{(9)(27)^2 + (9)(41)^2}{10 + 10 - 2}} = 34.713$$
$$t = \frac{807 - 757}{34.713\sqrt{\frac{1}{10} + \frac{1}{10}}} = 3.22$$

This indicates the P-value for a two-tailed test would fall between .002 and .01. The null hypothesis would be rejected at the level $\alpha = .01$. This indicates a large difference in compression strength as compared to normal variation which disagrees with the statement of the authors.

59 Summary quantities are $m = 24$, $\bar{x} = 103.66$, $s_1 = 3.74$, $n = 11$, $\bar{y} = 101.11$, $s_2 = 3.60$. We use the pooled t interval based on $24 + 11 - 2 = 33$ d.f.; 95% confidence requires $t_{.025,33} = 2.03$. With $s_p^2 = 13.68$ and $s_p = 3.70$, the CI is

$$2.55 \pm (2.03)(3.70)\left(\frac{1}{24} + \frac{1}{11}\right)^{1/2} = 2.55 \pm 2.73 = (-.18,\ 5.28).$$

61 Computing the difference as surface pH - subsoil pH, $\bar{d} = -0.0375$, $s_D = 0.2213$, $t_{0.05,7} = 1.895$

$$(-0.0375) \pm (1.895)\left(\frac{0.2213}{\sqrt{8}}\right)$$

(-.186, .111)

We assumed the distribution of the differences is normal.

Chapter 9

63

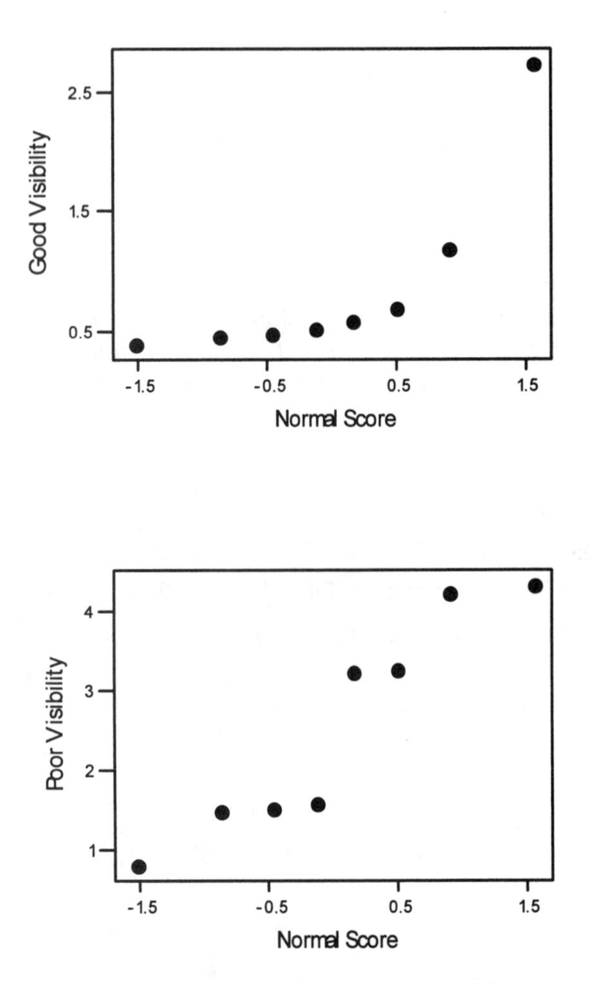

A normal probability plot indicates the data for good visibility does not follow a normal distribution, thus a t test is not appropriate for this small a sample size.

65 **a** With n denoting the second sample size, the first is $m = 3n$. We then wish

$$20 = 2(2.58)\left(\frac{900}{3n} + \frac{400}{n}\right)^{1/2},$$ which yields $n = 47$, $m = 141$.

b We wish to find the n that minimizes $2z_{\alpha/2}\left(\frac{900}{400-n}+\frac{400}{n}\right)^{1/2}$, or equivalently the n that minimizes $\frac{900}{400-n}+\frac{400}{n}$. Taking the derivative with respect to n and equating to 0 yields $900(400-n)^{-2}-400n^{-2}=0$, whence $9n^2=4(400-n)^2$, or $5n^2+3200n-640{,}000=0$. This yields $n=160$, $m=400-n=240$.

67 **a**

1. Parameter of interest: $\mu_1-\mu_2$ = true difference in insulin-binding capacity for rats treated with low and high doses of insulin.
2. H_0: $\mu_1-\mu_2=0$
3. H_a: $\mu_1-\mu_2\neq 0$
4. $t=\dfrac{(\bar{x}-\bar{y})-0}{s_p\sqrt{\frac{1}{m}+\frac{1}{n}}}$
5. RR: either $t\geq t_{.0005,18}$ or $t\leq -t_{.0005,18}$, i.e., either $t\geq 3.922$ or $t\leq -3.922$.
6. $s_p=\sqrt{\dfrac{(7)(.51)^2+(11)(.35)^2}{8+12-2}}=.4195$

   $t=\dfrac{(1.98-1.30)-0}{(.4195)\sqrt{\frac{1}{8}+\frac{1}{12}}}=3.551$
7. Fail to reject H_0. The difference in insulin-binding capacity is not statistically significant at the level $\alpha=.001$.

b $.002<P<.01$. Since $t=3.551$ falls between $t_{.005}$ and $t_{.001}$ and the test is two-tailed.

69 $\Delta_0=0$, $\sigma_1=\sigma_2=10$, $d=1$, $\sigma=\left(\frac{200}{n}\right)^{1/2}=\frac{14.142}{\sqrt{n}}$, so $\beta=\Phi\left(1.645-\frac{\sqrt{n}}{14.142}\right)$, giving $\beta=.9015$, $.8264$, $.0294$, and $.0000$ for $n=25$, 100, 2500, and 10,000 respectively. If the μ_i's referred to true average IQs resulting from two different conditions, $\mu_1-\mu_2=1$ would have little practical significance, yet very large sample sizes would yield statistical significance in this situation.

Chapter 9

71 H_0: $p_1 = p_2$ will be rejected at level α in favor of H_a: $p_1 \neq p_2$ if either $z \geq z_{.025} = 1.96$ or if $z \leq -1.96$. With $\hat{p}_1 = .3438$, $\hat{p}_2 = .3125$ and $\hat{p} = .3267$, $z = \dfrac{.0313}{.0502} = .62$, so H_0 cannot be rejected and there appears to be no difference.

73 The hypotheses of interest are H_0: $\mu_D = 0$ versus H_a: $\mu_D > 0$ with H_0 rejected at level .1 if $t \geq 1.350$. With $\bar{d} = .821$, $s_D = 2.52$, $t_{\text{paired}} = \dfrac{.821\sqrt{14}}{2.52} = 1.22$. Since 1.22 is not ≥ 1.350, H_0 is not rejected at level .10.

75 **a** Let μ_1 and μ_2 denote true average weights for operations 1 and 2, respectively. The hypotheses are H_0: $\mu_1 - \mu_2 = 0$ versus H_a: $\mu_1 - \mu_2 \neq 0$. Because both m and n are large, a two-sample z test (requiring no assumptions about the two weight distributions) will be used. H_0 will be rejected if either $z \geq 1.96$ or $z \leq -1.96$. We calculate

$$z = \frac{1402.24 - 1419.63}{[(10.97)^2/30 + (9.96)^2/30]^{1/2}} = \frac{-17.39}{2.705} = -6.4.$$

Because $-6.4 \leq -1.96$, H_0 should be rejected. The conclusion is that the true average weights differ.

b H_0: $\mu_1 = 1400$ will be tested against H_a: $\mu_1 > 1400$ using a one-sample z test with test statistic value $z = \dfrac{\bar{x} - 1400}{s_1/\sqrt{m}}$. H_0 should be rejected if $z \geq 1.645$.

The test statistic value is $\dfrac{1402.24 - 1400}{10.97/\sqrt{30}} = \dfrac{2.24}{2.00} = 1.1$. Because $1.1 < 1.645$, H_0 is not rejected. True average weight does not appear to exceed 1400.

77 $\text{Var}(\bar{X} - \bar{Y}) = \dfrac{\lambda_1}{m} + \dfrac{\lambda_2}{n}$ and $\hat{\lambda}_1 = \bar{X}$, $\hat{\lambda}_2 = \bar{Y}$, $\hat{\lambda} = \dfrac{m\bar{X} + n\bar{Y}}{m + n}$, giving $Z = \dfrac{\bar{X} - \bar{Y}}{(\hat{\lambda}/m + \hat{\lambda}/n)^{1/2}}$.

With $\bar{x} = 1.616$ and $\bar{y} = 2.557$, $z = -5.3$ and $P = 2(1 - \Phi(5.3)) < .0006$, so we would certainly reject H_0: $\lambda_1 = \lambda_2$ in favor of H_a: $\lambda_1 \neq \lambda_2$.

Chapter 9

79 The hypotheses to be tested are H_0: $\sigma_1^2 = \sigma_2^2$ versus H_a: $\sigma_1^2 \neq \sigma_2^2$, and $f = \frac{s_1^2}{s_2^2} = 4.57$. Since $F_{.05,5,4} = 6.26$, H_0 cannot be rejected at significance level .10.

CHAPTER 10

Section 10.1

1 H_o will be rejected if $f \geq F_{0.05,4,15} = 3.06$ (since $I-1 = 4$, $I(J-1) = (5)(3) = 15$). The computed value of F is $f = \frac{2573.3}{1394.2} = 1.85$. Since 1.85 is not ≥ 3.06, H_o is not rejected.

3 With μ_i = true average lumen output for brand i bulbs, we wish to test $H_0 : \mu_1 = \mu_2 = u_3$ vs. H_a : at least two μ_i 's are unequal.
$I-1 = 2$ and $I(J-1) = 21$, so H_o will be rejected at level 0.05 if $f \geq F_{0.05,2,21}$ $= 3.47$. $MSTr = \hat{\sigma}_B^2 = \frac{591.2}{2} = 295.60$, $MSE = \hat{\sigma}_W^2 = \frac{4773.3}{21} = 227.30$,

so $f = \frac{295.60}{227.30} = 1.30$. Since 1.30 is not ≥ 3.47, H_0 is not rejected.

5 μ_i = true mean modulus of elasticity for grade $i (i = 1, 2, 3)$
H_0: $\mu_1 = \mu_2 = \mu_3$ H_a: at least two means differ
$F_{.01,2,27} = 5.49$ grand mean = 1.5367

$$MSTr = \frac{10}{2}[(1.63-1.5367)^2+(1.56-1.5367)^2+(1.42-1.5367)^2] = .1143$$

$$MSE = \frac{1}{3}[(.27)^2+(.24)^2+(.26)^2] = .0660$$

$$f = \frac{MSTr}{MSE} = \frac{.1143}{.0660} = 1.73$$

Fail to reject H_0. The three grades do not appear to differ.

7

Source	df	SS	MS	f
Treatments	3	75,081.72	25,027.24	1.70
Error	16	235,419.04	14,713.69	
Total	19	310,500.76		

The hypotheses are $H_0 : \mu_1 = \mu_2 = \mu_3 = \mu_4$ vs. H_a : at least two of the 4 μ_i's are unequal. $F_{.05,3,16} = 3.24$, and since 1.70 is not ≥ 3.24, H_0 is not rejected.

Chapter 10

9 $x_{1\cdot} = 34.3,\ x_{2\cdot} = 39.6,\ x_{3\cdot} = 33.0,\ x_{4\cdot} = 41.9,\ x_{\cdot\cdot} = 148.8,\ \Sigma\Sigma x_{ij}^2 = 946.68$

so $CF = \frac{(148.8)^2}{24} = 922.56,\ SST = 946.68 - 922.56 = 24.12,$

$SSTr = \frac{(34.3)^2 + ... + (41.9)^2}{6} - 922.56 = 8.98,\ \text{SSE} = 24.12 - 8.98 = 15.14.$

Source	df	SS	MS	f
Treatments	3	8.98	2.99	3.95
Error	20	15.14	.757	
Total	23	24.12		

$F_{0.05,3,20} = 3.10$ and $3.95 \geq 3.10$, so H_o is rejected at level .05.

Section 10.2

11 $Q_{0.05,5,15} = 4.37,\ w = 4.37\sqrt{\frac{272.8}{4}} = 36.09.$

3	1	4	2	5
437.5	462.0	469.3	512.8	532.1

The brands seem to divide into two groups - 1, 3, and 4, and 2 and 5 - with no significant differences within each group but all between group differences significant.

13

3	1	4	2	5
427.5	462.0	469.3	502.8	532.1

Brand 1 does not differ significantly from 3 or 4, 2 does not differ significantly from 4 or 5, 3 does not differ significantly from 1, 4 does not differ significantly from 1or 2, 5 does not differ significantly from 2, but all other differences (e.g. 1 with 2 and 4, 2 with 3, etc.) do appear to be significant.

Chapter 10

15 $Q_{0.01,4,36} \approx 4.75$, $w = 4.75\sqrt{\frac{15.64}{10}} = 5.94$.

2	1	3	4
24.69	26.08	29.95	33.84

(Underscoring: 24.69, 26.08, 29.95 underscored together; 29.95, 33.84 underscored together.)

This underscoring is identical in structure to that of exercise 14 above, so the conclusion is identical.

17 $\theta = \sum c_i \mu_i$ where $c_1 = c_2 = .5$ and $c_3 = -1$, so $\hat{\theta} = .5\bar{x}_{1\cdot} + .5\bar{x}_{2\cdot} - \bar{x}_{3\cdot} = -.396$ and $\sum c_i^2 = 1.50$. With $t_{.025,6} = 2.447$ and $MSE = .03106$, the CI is, from (10.5) on page 406, $-.396 \pm (2.447)\left[\frac{(.03106)(1.50)}{3}\right]^{1/2} = -.396 \pm .305 = (-.701, -.091)$

19 $MSTr = 140$, error d.f. = 12, so $f = \frac{140}{SSE/12} = \frac{1680}{SSE}$ and $F_{.05,2,12} = 3.89$.

$w = Q_{0.05,3,12}\sqrt{\frac{MSE}{J}} = 3.77\sqrt{\frac{SSE}{60}} = .4867\sqrt{SSE}$. Thus we wish $\frac{1680}{SSE} > 3.89$ (significance of f) and $.4867\sqrt{SSE} > 10$ (= 20 – 10, the difference between the extreme $\bar{x}_{i\cdot}$'s - so no significant differences are identified). These become 431.88 > SSE and SSE > 422.16, so SSE = 425 will work.

21

a

Grand mean = 222.167
$MSTr$ = 38,015.1333
MSE = 1,681.8333
f = 22.6
H_0: $\mu_1 = \mu_2 = \mu_3 = \mu_4 = \mu_5 = \mu_6$
H_a: at least two means differ.
$F_{.01,5,78}$ is not tabled so use $F_{.01,5,60} = 3.34$
Reject H_0. The data indicates there is a dependence on injection regimen.

b Assume $t_{.005,78} \approx 2.645$

i $\mu_1 - \frac{1}{5}(\mu_2 + \mu_3 + \mu_4 + \mu_5 + \mu_6)$

$$\sum c_i \bar{x}_i \pm t_{\alpha/2}\ I(J-1)\sqrt{\frac{MSE\sum c_i^2}{J}}$$

$$-67.4 \pm (2.645)\sqrt{\frac{1{,}681.8333(1.2)}{14}}$$

(-99.16, -35.64)

ii $\frac{1}{4}(\mu_2+\mu_3+\mu_4+\mu_5)-\mu_6$

$$61.75 \pm (2.645)\sqrt{\frac{1681.8333(1.25)}{14}}$$

(29.34, 94.16)

Section 10.3

23 $J_1 = 5$, $J_2 = 4$, $J_3 = 4$, $J_4 = 5$, $\bar{x}_{1\cdot} = 58.28$, $\bar{x}_{2\cdot} = 55.40$, $\bar{x}_{3\cdot} = 50.85$, $\bar{x}_{4\cdot} = 45.50$, $MSE = 8.89$

With $W_{ij} = Q_{.05,4,14} \cdot \sqrt{MSE} \cdot \sqrt{\frac{1}{2}\left(\frac{1}{J_i}+\frac{1}{J_j}\right)} = 12.25\sqrt{\frac{1}{2}\left(\frac{1}{J_i}+\frac{1}{J_j}\right)}$

$\bar{x}_{1\cdot} - \bar{x}_{2\cdot} \pm W_{12} = 2.88 \pm 5.81$

$\bar{x}_{1\cdot} - \bar{x}_{3\cdot} \pm W_{13} = 7.43 \pm 5.81$ *

$\bar{x}_{1\cdot} - \bar{x}_{4\cdot} \pm W_{14} = 12.78 \pm 5.48$ *

$\bar{x}_{2\cdot} - \bar{x}_{3\cdot} \pm W_{23} = 4.55 \pm 6.13$

$\bar{x}_{2\cdot} - \bar{x}_{4\cdot} \pm W_{24} = 9.90 \pm 5.81$ *

$\bar{x}_{3\cdot} - \bar{x}_{4\cdot} \pm W_{34} = 5.35 \pm 5.81$

* Identifies an interval that doesn't include zero, corresponding to μ's that are judged significantly different.

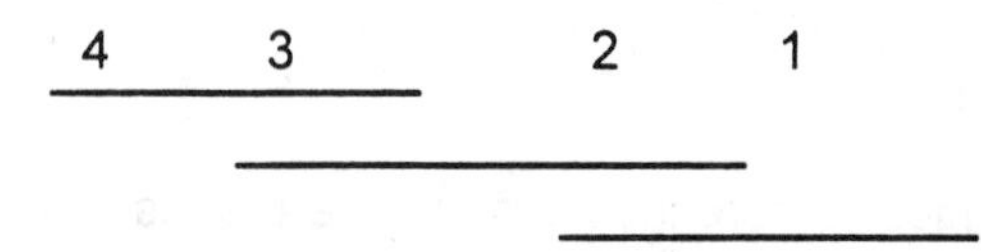

This underscoring pattern does not have a very straightforward interpretation.

Chapter 10

25

Diet	J_i	$\bar{x}_{i\cdot}$	$s_{i\cdot}$	$x_{i\cdot}$
Red Maple '74	13	1.134	.0252	14.742
Red Oak/Red Maple	10	1.148	.0253	11.480
Red Maple '75	20	1.159	.0179	23.180
Red Oak	16	1.191	.0200	19.056
Red Oak/White Pine	16	1.217	.0160	19.472
	75			87.93

where $X_{i\cdot} = J_i \cdot \bar{X}_{i\cdot}$

$$X_{i\cdot}^2 = (S_i^2)(J_i - 1) + \frac{(X_{i\cdot})^2}{J_i} \quad \text{(from computational formula for variance)}$$

a

$$\bar{x}_{\cdot\cdot} = \frac{87.93}{75} = 1.1724$$

$$SSTr = \frac{(14.742)^2}{13} + \frac{(11.480)^2}{10} + \frac{(23.180)^2}{20} + \frac{(19.056)^2}{16} + \frac{(19.472)^2}{16} - \frac{(87.93)^2}{75}$$

$$SSTr = .066076$$

$$MSTr = \frac{SSTr}{I-1} = \frac{.066076}{4} = .016519$$

b

$$SSE = (12)(.0252)^2 + (9)(.0253)^2 + (19)(.0179)^2 + (15)(.0200)^2 + (15)(.0160)^2$$
$$= .029309$$

$$MSE = \frac{SSE}{n-I} = \frac{.029309}{70} = .0004187$$

c

$F_{.05,4,70}$ not tabled so use $F_{.05,4,60} = 2.53$

$$f = \frac{MSTr}{MSE} = \frac{.016519}{.0004187} = 39.45$$

Reject H_0. The data does suggest differences among the means.

d

$$W_{ij} = Q_{\alpha,I,n-I} \cdot \sqrt{\frac{MSE}{2}\left(\frac{1}{J_i} + \frac{1}{J_j}\right)}$$

$Q_{.05,5,70}$ not tabled so use $Q_{.05,5,60} = 3.98$

w_{ij}	$\bar{x}_j - \bar{x}_i$	
w_{12} = .024	.014	not significantly different
w_{13} = .02	.025	significantly different
w_{23} = .02	.011	not significantly different
w_{24} = .023	.048	significantly different
w_{34} = .019	.032	significantly different
w_{45} = .020	.026	significantly different

1 2 3 4 5

(1 and 2 underscored together; 2 and 3 underscored together)

27 Let μ_i denote the true average folacin content for specimens of brand i. The hypotheses to be tested are $H_0: \mu_1 = \mu_2 = \mu_3 = \mu_4$ vs H_a: at least two μ_i's differ. $I-1 = 3$ and $n-I = 24-4 = 20$, so H_0 will be rejected at level 0.05 if $f \geq 3.10$.

$\Sigma\Sigma x_{ij}^2 = 1246.88$ and $\frac{x_{..}^2}{n} = \frac{(168.4)^2}{24} = 1181.61$, so $SST = 65.27$.

$$\frac{\Sigma x_{i.}^2}{J_i} = \frac{(57.9)^2}{7} + \frac{(37.5)^2}{5} + \frac{(38.1)^2}{6} + \frac{(34.9)^2}{6} = 1205.10,$$

so $SSTr = 1205.10 - 1181.61 = 23.49$.

Source	df	SS	MS	f
Treatments	3	23.49	7.83	3.75
Error	20	41.78	2.09	
Total	23	65.27		

$3.75 \geq 3.10$, so reject H_0 in favor of H_a.

29
$$E(SSTr) = E[\Sigma J_i \bar{X}_{i.}^2 - n\bar{X}_{..}^2] = \Sigma J_i E(\bar{X}_{i.}^2) - nE(\bar{X}_{..}^2)$$
$$= \Sigma J_i[\text{Var}(\bar{X}_{i.}) + (E(\bar{X}_{i.}))^2] - n[\text{Var}(\bar{X}_{..}) + (E(\bar{X}_{..}))^2]$$
$$= \Sigma J_i\left[\frac{\sigma^2}{J_i} + \mu_i^2\right] - n\left[\frac{\sigma^2}{n} + \frac{(\Sigma J_i \mu_i)^2}{n}\right]$$

$= (I-1)\,\sigma^2 + \Sigma J_i(\mu+\alpha_i)^2 - [\Sigma J_i(\mu+\alpha_i)]^2$

$= (I-1)\,\sigma^2 + \Sigma J_i\mu^2 + 2\mu\Sigma J_i\alpha_i + \Sigma J_i\alpha_i^2 - [\mu\Sigma J_i]^2$

$= (I-1)\,\sigma^2 + \Sigma J_i\alpha_i^2$, from which $E(MSTr)$ is obtained through division by $I - 1$.

31 With $\sigma = 1$ (any other σ will yield the same ϕ), $\alpha_1 = -1$, $\alpha_2 = \alpha_3 = 0$, $\alpha_4 = 1$,

$$\Phi^2 = \frac{0.25[5(-1)^2+4(0)^2+4(0)^2+5(1)^2]}{1} = 2.5,\ \Phi = 1.58,\ \nu_1 = 3,\ \nu_2 = 14,$$

power ≈ 0.62.

33 $g(x) = x\left(1-\frac{x}{n}\right) = nu(1-u)$ where $u = \frac{x}{n}$, so $h(x) = \int [u(1-u)]^{-1/2}\,du$.

From a table of integrals, this gives $h(x) = \arcsin(\sqrt{u}) = \arcsin\left(\sqrt{\frac{x}{n}}\right)$ as the appropriate transformation.

Supplementary

35 **a** H_0: $\mu_1 = \mu_2 = \mu_3 = \mu_4$

H_a: at least two means differ

$F_{0.01,3,20} = 4.94$ thus fail to reject H_0. The means do not appear to differ.

b At the level $\alpha = .01$ you would still fail to reject H_0 because P-value > alpha.

37 $SSE = 17(18.71)^2 + 27(21.15)^2 + 18(17.87)^2 = 23{,}776.86$,

$$SSTr = \frac{(7617.78)^2}{18} + \frac{(11{,}707.64)^2}{28} + \frac{(8004.13)^2}{19} - \frac{(27{,}329.55)^2}{65} = 300.60,$$

so $MSE = 383.5$, $MSTr = 150.3$, $f = .39$. This f is clearly insignificant, so there appears to be no differences.

39 $\hat{\theta} = 2.58 - \frac{2.63+2.13+2.41+2.49}{4} = .165$, $t_{.025,25} = 2.060$, $MSE = .108$,

and $\Sigma c_i^2 = (1)^2 + (-.25)^2 + (-.25)^2 + (-.25)^2 + (-.25)^2 = 1.25$, so a 95% confidence interval for θ is $.165 \pm 2.060\left[\frac{(.108)(1.25)}{6}\right]^{1/2} = .165 \pm .309 = (-.144, .474)$. This interval does include zero, so 0 is a plausible value for θ.

Chapter 10

41 This is a random effects situation. $H_0 : \sigma_A^2 = 0$ states that variation in laboratories doesn't contribute to variation in percentage. H_0 will be rejected in favor of H_a if $f \geq F_{0.05,3,8} = 4.07$. $SST = 86{,}078.9897 - 86{,}077.2224 = 1.7673$, $SSTr = 1.0559$, and $SSE = .7114$. Thus $f = \dfrac{1.0559/3}{.7114/8} = 3.96$.
Because 3.96 is not ≥ 4.07, H_0 cannot be rejected at level .05. Variation in laboratories does not appear to be present.

43

Source	df	SS	MS	f	$F_{.05}$
Treatments	3	24,937.63	8312.54	1117.8	4.07
Error	8	59.49	7.44		
Total	11	24,997.12			

Because 1117.8 ≥ 4.07, $H_0 : \mu_1 = \mu_2 = \mu_3 = \mu_4$ is rejected.
$Q_{.05,4,8} = 4.53$, so $w = 4.53\left(\dfrac{7.44}{3}\right)^{1/2} = 7.13$. The four sample means are $\bar{x}_{4\cdot} = 29.92$, $\bar{x}_{1\cdot} = 33.96$, $\bar{x}_{3\cdot} = 115.84$, and $\bar{x}_{2\cdot} = 129.30$. Only $\bar{x}_{1\cdot} - \bar{x}_{4\cdot} < 7.13$, so all μ_i's are judged significantly different from one another except for μ_4 and μ_1 (corresponding to PCM and OCM).

45 The ordered residuals are -6.67, -5.67, -4, -2.67, -1, -1, 0, 0, 0, .33, .33, .33, 1, 1, 2.33, 4, 5.33, 6.33. The corresponding z percentiles are -1.91, -1.38, -1.09, -.86, -.67, -.51, -.36, -.21, -.07, .07, .21, .36, .51, .67, .86, 1.09, 1.38, and 1.91. The resulting plot (of (-1.91, -6.67), ..., (1.91, 6.33)) is reasonably straight.

CHAPTER 11

Section 11.1

1 **a** $MSA = \frac{30.6}{4} = 7.65$, $MSE = \frac{59.2}{12} = 4.93$, $f_A = \frac{7.65}{4.93} = 1.55$.
Since $F_{.05,4,12} = 3.26$ and 1.55 is not ≥ 3.26, don't reject H_{0A}.

b $MSB = \frac{44.1}{3} = 14.70$, $f_B = \frac{14.70}{4.93} = 2.98$. Since $F_{.05,3,12} = 3.49$ and 2.98 is not ≥ 3.49, do not reject H_{0B}.

3 $x_{1\bullet} = 927$, $x_{2\bullet} = 1301$, $x_{3\bullet} = 1764$, $x_{4\bullet} = 2453$, $x_{\bullet 1} = 1347$, $x_{\bullet 2} = 1529$, $x_{\bullet 3} = 1677$, $x_{\bullet 4} = 1892$, $x_{\bullet\bullet} = 6445$, $CF = \frac{(6445)^2}{16} = 2{,}596{,}126.56$, $\Sigma\Sigma x_{ij}^2 = 2{,}969{,}375$,
$SSA = 324{,}082.2$, $SSB = 39{,}934.2$, $SST = 373{,}248.4$, $SSE = 9232.0$.

a

Source	df	SS	MS	f
A	3	324,082.2	108,027.4	105.3
B	3	39,934.2	13,311.4	13.0
Error	9	9232.0	1025.8	
Total	15	373,248.4		

Since $F_{.01,3,9} = 6.99$, both H_{0A} and H_{0B} are rejected.

b $Q_{.01,4,9} = 5.96$, $w = 5.96\sqrt{1025.8/4} = 95.44$

i:	1	2	3	4
$\bar{x}_{i.}$:	231.75	325.25	441.00	613.25

(underline spans levels 1 and 2)

All levels of factor A (gas rate) differ significantly except for 1 and 2.

c

j:	1	2	3	4	$w = 95.4$
$\bar{x}_{.j}$:	336.75	382.25	419.25	473	

(underlines span levels 1–3 and levels 2–4)

Only levels 1 and 4 appear to differ significantly.

Chapter 11

5

Source	df	SS	MS	f
Angle	3	58.16	19.3867	2.5565
Connector	4	246.97	61.7425	8.1419
Error	12	91.00	7.5833	
Total	19	396.13		

H_0: $\alpha_1 = \alpha_2 = \alpha_3 = \alpha_4 = 0$ H_a: at least one α_i is not zero.
$f_A = 2.5565$, $F_{.01,3,12} = 5.95$
Fail to reject H_0. The data fails to indicate any effect due to the angle of pull.

7 a $CF = 140{,}454$, $SST = 3476$, $SSTr = [(905)^2 + (913)^2 + (936)^2]/18 - 140{,}454 = 28.78$.

$SSBl = \dfrac{430{,}295}{3} - 140{,}454 = 2977.67$, $SSE = 469.55$, $MSTr = 14.39$,

$MSE = 13.81$, $f_{Tr} = 1.04$, f_{Tr} is clearly insignificant when compared to $F_{.05,2,51}$.

b $f_{Bl} = 12.68$, which is significant and suggests substantial variation among subjects. If we had not controlled for such variation, it might have affected the analysis and conclusions.

9

Source	df	SS	MS	f
Treatment	3	81.1944	27.0648	22.36
Blocks	8	66.50	8.3125	6.87
Error	24	29.0556	1.2106	
Total	35	176.750		

$F_{.05,3,24} = 3.01$
Reject H_0. There is an effect due to treatments.

1	4	3	2
8.56	9.22	10.78	12.44

(8.56 and 9.22 underlined together)

$Q_{.05,4,24} = 3.90$

$$w = (3.90)\sqrt{\frac{1.2106}{9}} = 1.43$$

Chapter 11

11 **a** With $Y_{ij} = X_{ij} + d$, $\bar{Y}_{i.} = \bar{X}_{i.} + d$, $\bar{Y}_{.j} = \bar{X}_{.j} + d$, $\bar{Y}_{..} = \bar{X}_{..} + d$, so all quantities inside the parentheses in (11.5) remain unchanged when the Y quantities are substituted for the corresponding X's (e.g. $\bar{Y}_{i.} - \bar{Y}_{..} = \bar{X}_{i.} - \bar{X}_{..}$, etc.).

b With $Y_{ij} = cX_{ij}$, each sum of squares for Y is the corresponding SS for X multiplied by c^2. However, when F ratios are formed the c^2 factors cancel, so all F ratios computed from Y are identical to those computed from X. If $Y_{ij} = cX_{ij} + d$, the conclusions reached from using the Y's will be identical to those reached using the X's.

13 **a** $\Sigma\alpha_i^2 = 24$, so $\Phi^2 = \left(\frac{3}{4}\right)\left(\frac{24}{16}\right)$ 1.125, $\Phi = 1.06$, $\nu_1 = 3$, $\nu_2 = 6$, and from figure 10.5, power $\approx .2$. For the second alternative, $\Phi = 1.59$ and power $\approx .43$.

b $\Phi^2 = \left(\frac{I}{J}\right)\Sigma\beta_j^2/\sigma^2 = \left(\frac{4}{5}\right)\left(\frac{20}{16}\right) = 1.00$, so $\Phi = 1.00$, $\nu_1 = 4$, $\nu_2 = 12$, and power $\approx .3$.

Section 11.2

15 **a**

Source	df	SS	MS	f
Sand	2	705	352.5	3.76
Fiber	2	1,278	639.0	6.82*
Sand & Fiber	4	279	69.75	0.74
Error	9	843	93.67	
Total	17	3,105		

$F_{.05,2,9} = 4.26$ $F_{.05,4,9} = 3.63$

There appears to be an effect due to carbon fiber addition.

b

Source	df	SS	MS	f
Sand	2	106.78	53.39	6.54*
Fiber	2	87.11	43.56	5.33*
Sand & Fiber	4	8.89	2.22	.27
Error	9	73.50	8.17	
Total	17	276.28		

$F_{.05,2,9} = 4.26$ $F_{.05,4,9} = 3.63$

There appears to be an effect due to both sand and carbon fiber addition to casting hardness.

c

Sand%	Fiber%	$\bar{x}$
0	0	62
15	0	68
30	0	69.5
0	.25	69
15	.25	71.5
30	.25	73
0	.50	68
15	.50	71.5
30	.50	74

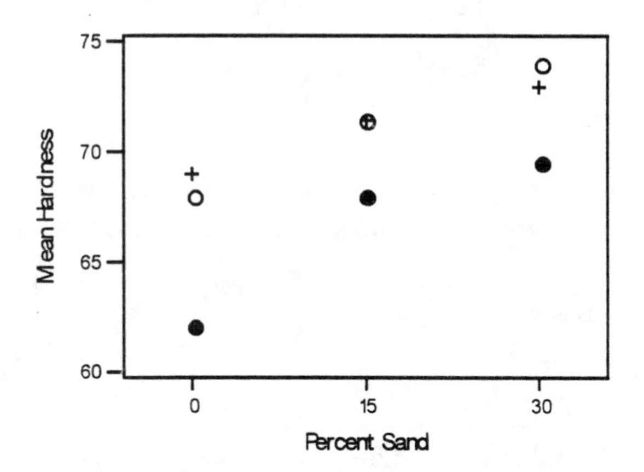

Solid circle = Fiber is 0%
Plus = Fiber is .25%
Open circle = Fiber is .50%

The plot indicates some effect due to sand and fiber addition with no significant interaction. This agrees with the statistical analysis in part b.

Chapter 11

17 a

	$x_{ij.}$	1	2	3	$x_{i..}$
	1	16.44	17.27	16.10	49.81
i	2	16.24	17.00	15.91	49.15
	3	16.80	17.37	16.20	50.37
	$x_{.j.}$	49.48	51.64	48.21	$x_{...} = 149.33$ $CF = 1238.8583$

(column index: j)

Thus $SST = 1240.1525 - 1238.8583 = 1.2942$, $SSE = 1240.1525 - \dfrac{2479.9991}{2}$ $= .1530$, $SSA = [(49.81)^2 + (49.15)^2 + (50.37)^2]/6 - 1238.8583 = .1243$, $SSB = 1.0024$.

Source	df	SS	MS	f
A	2	.12431	.06216	3.66
B	2	1.00241	.50121	29.49
AB	4	.01456	.00364	.21
Error	9	.15295	.01699	
Total	17	1.29423		

$F_{.01,4,9} = 6.42$, so H_{0AB} cannot be rejected. $F_{.01,2,9} = 8.02$, so H_{0A} cannot be rejected but we do reject H_{0B}; type of coal does appear to affect total acidity.

b $Q_{.01,3,9} = 5.43$, $w = 5.43\left(\dfrac{.0170}{6}\right)^{1/2} = .289$.

j:	3	1	2
$\bar{x}_{.j.}$:	8.035	8.247	8.607

(3 and 1 underlined together)

Coal 2 is judged significantly different from both 1 and 3, but these latter two don't differ significantly from one another.

Chapter 11

19 **a**

$$SST = 12{,}280{,}103 - \frac{(19{,}143)^2}{30} = 64{,}954.70,$$

$$SSE = 12{,}280{,}103 - \frac{24{,}529{,}699}{2} = 15{,}253.50$$

$$SSA = \frac{122{,}380{,}901}{10} - \frac{(19{,}143)^2}{30} = 22{,}941.80$$

$$SSB = 22{,}765.53$$

$$SSAB = 64{,}954.70 - [22{,}941.80 + 22{,}765.53 + 15{,}253.50] = 3993.87.$$

Source	df	SS	MS	f
A	2	22,941.80	11,470.90	$\frac{11{,}470.90}{499.23} = 22.98$
B	4	22,765.53	5691.38	$\frac{5691.38}{1016.9} = 5.60$
AB	8	3993.87	499.23	.49
Error	15	15,253.50	1016.90	
Total	29	64,954.70		

b $f_{AB} = .49$ is clearly not significant. Since $F_{.05,2,8} = 4.46$ and $22.98 \geq 4.46$, H_{0A} is rejected. Since $F_{.05,4,15} = 3.06$ and $5.60 \geq 3.06$, H_{0B} is rejected. Thus we conclude that the different cement factors affect flexural strength differently and that batch variability contributes to variation in flexural strength.

21

$$E(\bar{X}_{i..} - \bar{X}_{...}) = \frac{1}{JK}\sum_j\sum_k E(X_{ijk}) - \frac{1}{IJK}\sum_i\sum_j\sum_k E(X_{ijk}) = \frac{1}{JK}\sum_j\sum_k(\mu + \alpha_i + \beta_j + \gamma_{ij})$$

$$- \frac{1}{IJK}\sum_i\sum_j\sum_k(\mu + \alpha_i + \beta_j + \gamma_{ij}) = \mu + \alpha_i - \mu = \alpha_i.$$

23 With $\theta = \alpha_i - \alpha_{i'}$, $\hat{\theta} = \bar{X}_{i..} - \bar{X}_{i'..} = \frac{1}{JK}\sum_j\sum_k(X_{ijk} - X_{i'jk})$, and since $i \neq i'$ X_{ijk} and $X_{i'jk}$ are independent for every j, k. Thus $Var(\hat{\theta}) = Var(\bar{X}_{i..}) + Var(\bar{X}_{i'..}) = \frac{\sigma^2}{JK} + \frac{\sigma^2}{JK} = \frac{2\sigma^2}{JK}$ (because $Var(\bar{X}_{i..}) = Var(\bar{\varepsilon}_{i..})$ and $Var(\varepsilon_{ijk}) = \sigma^2$), so $\hat{\sigma}_{\hat{\theta}} = \sqrt{\frac{2MSE}{JK}}$. The appropriate number of df is $IJ(K-1)$, so the CI is $\bar{x}_{i..} - \bar{x}_{i'..} + t_{\alpha/2,IJ(K-1)}\sqrt{\frac{2MSE}{JK}}$. For the data of exercise 17, $\bar{x}_{2..} = 49.15$, $\bar{x}_{3..} = 50.37$, $MSE = .0170$, $t_{.025,9} = 2.262$, $J = 3$, $K = 2$, so the CI is $49.15 - 50.37 \pm (2.262)\sqrt{\frac{.0340}{6}} = -1.22 \pm .17 = (-1.39, -1.05)$.

Section 11.3

25 **a**

Source	df	SS	MS	f	$F_{0.05}$
A	2	14,144.44	7072.22	61.06	3.35
B	2	5,511.27	2,755.64	23.79	3.35
C	2	244,696.39	122,348.20	1,056.24	3.35
AB	4	1,069.62	267.41	2.31	2.73
AC	4	62.67	15.67	.14	2.73
BC	4	331.67	82.92	.72	2.73
ABC	8	1,080.77	135.10	1.17	2.31
Error	27	3,127.50	115.83		
Total	53	270,024.33			

b The computed F statistics for all four interaction terms are less than the tabled values for statistical significance at the level $\alpha = .05$. This indicates that none of the interactions is statistically significant.

c The computed F statistics for all three main effects exceed the tabled value for significance at the level $\alpha = .05$. All three main effects are statistically significant.

d $Q_{.05,3,27}$ is not tabled, use $Q_{.05,3,24} = 3.53$, $w = 3.53\left(\sqrt{\frac{115.83}{3 \cdot 3 \cdot 2}}\right) = 8.95$

All three levels differ significantly from each other.

27 $I = 3 \quad J = 2 \quad K = 4 \quad L = 4$

$SSA = JKL \sum (\bar{x}_{i...} - \bar{x}_{....})^2$; $SSB = IKL \sum (\bar{x}_{.j..} - \bar{x}_{....})^2$; $SSC = IJL \sum (\bar{x}_{..k.} - \bar{x}_{....})^2$

for level A: $\bar{x}_{1...} = 3.781 \quad \bar{x}_{2...} = 3.625 \quad \bar{x}_{3...} = 4.469$

for level B: $\bar{x}_{.1..} = 4.979 \quad \bar{x}_{.2..} = 2.938$

for level C: $\bar{x}_{..1.} = 3.417 \quad \bar{x}_{..2.} = 5.875 \quad \bar{x}_{..3.} = .875 \quad \bar{x}_{..4.} = 5.667$

$\bar{x}_{....} = 3.958$

$SSA = 12.907$; $SSB = 99.976$; $SSC = 393.436$

a

Source	df	SS	MS	f	Use 60 df for denominator $F_{.05}$
A	2	12.896	6.448	1.04	3.15
B	1	100.041	100.041	16.10	4.00
C	3	393.416	131.139	21.10	2.76
AB	2	1.646	.823	< 1	3.15
AC	6	71.021	11.837	1.905	2.25
BC	3	1.542	.514	< 1	2.76
ABC	6	9.771	1.629	< 1	2.25
Error	72	447.500	6.215		
Total	95	1,037.833			

b No interaction effects are significant at the level $\alpha = .05$.

c Factor B and C main effects are significant at the level $\alpha = .05$.

d $Q_{.05,4,72}$ not tabled so use $Q_{.05,4,60} = \mathbf{3.74}$,

$$w = Q \cdot \sqrt{\frac{MSE}{3 \cdot 2 \cdot 4}} = (3.74)\sqrt{\frac{6.215}{24}} = 1.90$$

machine:	3	1	4	2
mean:	.875	3.417	5.667	5.875

(5.667 and 5.875 underlined together)

29

$x_{ij.}$	B_1	B_2	B_3
A_1	210.2	224.9	218.1
A_2	224.1	229.5	221.5
A_3	217.7	230.0	202.0
$x_{.j.}$	652.0	684.4	641.6

$x_{i.k}$	A_1	A_2	A_3
C_1	213.8	222.0	205.0
C_2	225.6	226.5	223.5
C_3	213.8	226.6	221.2
$x_{i..}$	653.2	675.1	649.7

Chapter 11

$x_{.jk}$	C_1	C_2	C_3
B_1	213.5	220.5	218.0
B_2	214.3	246.1	224.0
B_3	213.0	209.0	219.6
$x_{..k}$	640.8	675.6	661.6

$\Sigma\Sigma x_{ij.}^2 = \mathbf{435{,}382.26}$ $\Sigma\Sigma x_{i.k}^2 = \mathbf{435{,}156.74}$ $\Sigma\Sigma x_{.jk}^2 = \mathbf{435{,}666.36}$

$\Sigma x_{.j.}^2 = \mathbf{1{,}305{,}157.92}$ $\Sigma x_{i..}^2 = \mathbf{1{,}304{,}540.34}$ $\Sigma x_{..k}^2 = \mathbf{1{,}304{,}774.56}$

Also $\Sigma\Sigma\Sigma x_{ijk}^2 = \mathbf{145{,}386.40}$, $x_{...} = \mathbf{1978}$, $CF = \mathbf{144{,}906.81}$, from which we obtain the ANOVA table displayed in the problem statement. $F_{.01,4,8} = 7.01$, so the *AB* and *BC* interactions are significant (as can also be seen from the *P*-values) and tests for main effects are not appropriate.

31

Source	df	SS	MS	f
A	6	67.32	11.02	
B	6	51.06	8.51	
C	6	5.43	.91	.61
Error	30	44.26	1.48	
Total	48	168.07		

$F_{.05,\,6,\,30} = 2.42$; since $.61 < 2.42$, treatment was not effective.

33

	1	2	3	4	5	
$x_{i..}$:	40.68	30.04	44.02	32.14	33.21	$\Sigma x_{i..}^2 = \mathbf{6630.91}$
$x_{.j.}$:	29.19	31.61	37.31	40.16	41.82	$\Sigma x_{.j.}^2 = \mathbf{6605.02}$
$x_{..k}$:	36.59	36.67	36.03	34.50	36.30	$\Sigma x_{..k}^2 = \mathbf{6489.62}$

$x_{...} = 180.09$, $CF = 1297.30$, $\Sigma\Sigma x_{ij(k)}^2 = \mathbf{1358.60}$. The ANOVA table and conclusions appear in the answer section of the textbook.

35

Source	df	MS	f	$^*F_{0.01}$
A	2	2207.329	2259.29	5.39
B	1	47.255	48.37	7.56
C	2	491.783	503.36	5.39
D	1	.044	.05	7.56
AB	2	15.303	15.66	5.39
AC	4	275.446	281.93	4.02
AD	2	.470	.48	5.39
BC	2	2.141	2.19	5.39
BD	1	.273	.28	7.56
CD	2	.247	.25	5.39
ABC	4	3.714	3.80	4.02
ABD	2	4.072	4.17	5.39
ACD	4	.767	.79	4.02
BCD	2	.280	.29	5.39
ABCD	4	.347	.355	4.02
Error	36	.977		
Total	71	93.621		

$SS_{total} = (71)(93.621) = 6{,}647.091$

Computing all other sums of squares and adding them up = 6,645.702.

Thus $SS_{ABCD} = 6{,}647.091 - 6{,}645.702 = 1.389$ and $MS_{ABCD} = \frac{1.116}{4} = .347$

*Note: Because a denominator degrees of freedom for 36 is not tabled, use 30. At the level $\alpha = .01$ the statistically significant main effects are A, B, and C. The interaction AB and AC are also statistically significant. No other interactions are statistically significant.

Chapter 11

Section 11.4

37

Condition	Total	1	2	Contrast	SS=(Contrast)^2/24
111	315	927	2478	5485	
211	612	1551	3007	1307	A = 71,177.04
121	584	1163	680	1305	B = 70,959.38
221	967	1844	627	199	AB = 1650.04
112	453	297	624	529	C = 11,660.04
212	710	383	681	-53	AC = 117.04
122	737	257	86	57	BC = 135.38
222	1107	370	113	27	ABC = 30.38

a $\hat{\beta}_1 = \bar{x}_{.2..} - \bar{x}_{....} = \dfrac{584+967+737+1107-315-612-453-710}{24} = 54.38.$

$\hat{\gamma}^{AC}_{11} = \dfrac{315-612+584-967-453+710-737+1107}{24} = -2.21.$

$\hat{\gamma}^{AC}_{21} = -\hat{\gamma}^{AC}_{11} = 2.21.$

b Factor SS's appear above. With $CF = \dfrac{5485^2}{24} = 1{,}253{,}551.04$ and

$\Sigma\Sigma\Sigma\Sigma x^2_{ijkl} = 1{,}411{,}889$, SST = 158,337.96, from which SSE = 2608.7. The ANOVA table appears in the answer section.
$F_{.05,1,16} = 4.49$, from which we see that the AB interaction and all main effects are significant.

39 $\Sigma\Sigma\Sigma\Sigma\Sigma x^2_{ijklm} = 3{,}308{,}143$, $x_{.....} = 11{,}959$, so $CF = \dfrac{11{,}959^2}{48} = 2{,}979{,}535.02$ and SST = 328,607.98. Each SS is (effect contrast)^2/48 and SSE is obtained by subtraction. The ANOVA table appears in the answer section. $F_{.05,1,32} \approx 4.15$, a value exceeded by the F ratios for AB interaction and the four main effects.

Chapter 11

41

Condition/Effect	SS = (constrast)^2/16	*f*
(1)	-	
A	.436	< 1
B	.099	< 1
AB	.003	< 1
C	.109	< 1
AC	.078	< 1
BC	1.404	3.62
ABC	.051	-
D	414.12	851
AD	.017	< 1
BD	.456	< 1
ABD	.055	-
CD	2.190	4.50
ACD	1.020	-
BCD	.133	-
ABCD	.681	-

$SSE = .051 + .055 + 1.020 + .133 + .681 = 1.940$, $df. = 5 \Rightarrow MSE = .388$
$F_{.05,1,5} = 6.61$, so only the D main effect is significant.

43 **a** The allocation of treatments to blocks is as given in the answer section, with block #1 containing all treatments having an even number of letters in common with both ab and cd, etc.

b $x_{.....} = 16{,}898$, so $SST = \mathbf{9{,}035{,}054} - \dfrac{\mathbf{16{,}898^2}}{\mathbf{32}} = \mathbf{111{,}853.88}.$

The eight block × replication totals are 2091 (= $\mathbf{618} + \mathbf{421} + \mathbf{603} + \mathbf{449}$, the sum of the four observations in block #1 on replication #1), 2092, 2133, 2145, 2113, 2080, 2080, 2122, and 2122, so $\boldsymbol{SSB1} = \dfrac{\mathbf{2091^2}}{\mathbf{4}} + \ldots + \dfrac{\mathbf{2122^2}}{\mathbf{4}} - \dfrac{\mathbf{16{,}898^2}}{\mathbf{32}} = \mathbf{898.88}.$

The remaining *SS*'s as well as all *F* ratios appear in the ANOVA table in the answer section. With $F_{.01,1,12} = 9.33$, only the A and B main effects are significant.

45 See the answer section.

47

		A	B	C	D	E	AB	AC	AD	AE	BC	BD	BE	CD	CE	DE
a	70.4	+	-	-	-	-	-	-	-	-	+	+	+	+	+	+
b	72.1	-	+	-	-	-	-	+	+	+	-	-	-	+	+	+
c	70.4	-	-	+	-	-	+	-	+	+	-	+	+	-	-	+
abc	73.8	+	+	+	-	-	+	+	-	-	+	-	-	-	-	+
d	67.4	-	-	-	+	-	+	+	-	+	+	-	+	-	+	-
abd	67.0	+	+	-	+	-	+	-	+	-	-	+	-	-	+	-
acd	66.6	+	-	+	+	-	-	+	+	-	-	-	+	+	-	-
bcd	66.8	-	+	+	+	-	-	-	-	+	+	+	-	+	-	-
e	68.0	-	-	-	-	+	+	+	+	-	+	+	-	+	-	-
abe	67.8	+	+	-	-	+	+	-	-	+	-	-	+	+	-	-
ace	67.5	+	-	+	-	+	-	+	-	+	-	+	-	-	+	-
bce	70.3	-	+	+	-	+	-	-	+	-	+	-	+	-	+	-
ade	64.0	+	-	-	+	+	-	-	+	+	+	-	-	-	-	+
bde	67.9	-	+	-	+	+	-	+	-	-	-	+	+	-	-	+
cde	65.9	-	-	+	+	+	+	-	-	-	-	-	-	+	+	+
abcde	68.0	+	+	+	+	+	+	+	+	+	+	+	+	+	+	+

Thus $SSA = \dfrac{(70.4 - 72.1 - 70.4 + \ldots + 68.0)^2}{16} = 2.250$, $SSB = 7.840$, $SSC = .360$, $SSD = 52.563$, $SSE = 10.240$, $SSAB = 1.563$, $SSAC = 7.563$, $SSAD = .090$, $SSAE = 4.203$, $SSBC = 2.103$, $SSBD = .010$, $SSBE = .123$, $SSCD = .010$, $SSCE = .063$, $SSDE = 4.840$, Error SS = sum of two factor SS's = 20.568, Error $MS = 2.057$, $F_{.01,1,10} = 10.04$, so only the D main effect is significant.

49

Source	df	SS	MS	f
A	1	322.667	322.667	980.38
B	3	35.623	11.874	36.08
AB	3	8.557	2.852	8.67
Error	16	5.266	.329	
Total	23	372.113		

We first test the null hypothesis of no interactions (H_0: $\gamma_{ij} = 0$ for all i, j). H_0 will be rejected at level .05 if $f_{AB} = \dfrac{MSAB}{MSE} \geq F_{.05,3,16} = 3.24$. Because $8.67 \geq 3.24$, H_0 is rejected. Because we have concluded that interaction is present, test for main effects are not appropriate.

51

Source	df	SS	MS	f
A (pressure)	1	6.94	6.940	11.57
B (time)	3	5.61	1.870	3.12
C (concen.)	2	12.33	6.165	10.28
AB	3	4.05	1.350	2.25
AC	2	7.32	3.660	6.10
BC	6	15.80	2.633	4.39
ABC	6	4.37	.728	1.21
Error	24	14.40	.600	
Total	47	70.82		

$F_{.05,6,24} = 2.51$, so there appear to be no 3-factor interactions. However, both AC and BC 2-factor interactions appear to be present.

Chapter 11

53

Source	df	SS	MS	f
A (diet)	2	18138	9069	28.9
B (temp.)	2	5182	2591	8.3
Interaction	4	1737	434.3	1.4
Error	36	11291	313.6	
Total	44	36348		

$F_{.05,4,36} = 2.63$ and $1.4 < 2.63$, so interaction appears to be absent. However, both 28.9 and 8.3 exceed the corresponding F critical values, so both diet and temperature appear to affect expected energy intake.

55

$\hat{\mu} = 402.8125$

$\hat{\alpha}_1 = -171.0625$ $\qquad \hat{\beta}_1 = -66.0625$

$\hat{\alpha}_2 = -77.5625$ $\qquad \hat{\beta}_2 = -20.5625$

$\hat{\alpha}_3 = 38.1875$ $\qquad \hat{\beta}_3 = 16.4375$

$\hat{\alpha}_4 = 210.4375$ $\qquad \hat{\beta}_4 = 70.1875$

Residuals (Ranked)	Percentile	z-percentile
-47.0	3.125	-1.87
-40.9	9.375	-1.32
-17.7	15.625	-1.01
-14.4	21.875	-.78
-11.7	28.125	-.58
-8.2	34.375	-.40
-5.9	40.625	-.24
-4.4	46.875	-.08
4.6	53.125	.08
5.8	59.375	.24
7.3	65.625	.40
14.8	71.875	.58
15.3	78.125	.78
18.8	84.375	1.01
34.3	90.625	1.32
49.6	96.875	1.87

Chapter 11

Based on a normal probability plot of the residuals, it seems reasonable to assume that they follow a normal distribution.

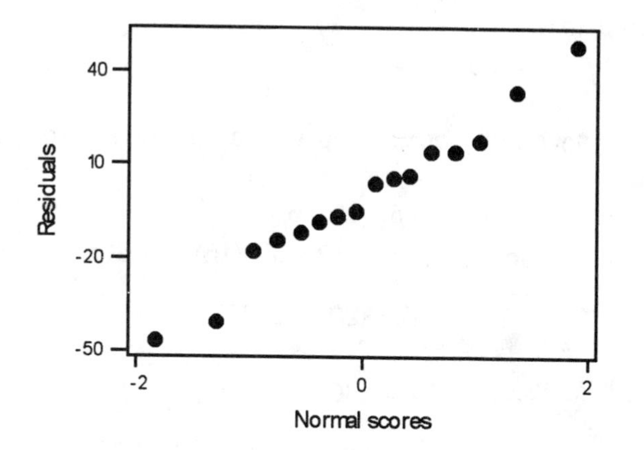

CHAPTER 12

Section 12.1

1 **a** $\mu_{Y \bullet 2500} = 1800 + 1.3(2500) = 5050$

b expected change = slope = β_1 = 1.3

c expected change = $100\,\beta_1$ = 130

d expected change = $-100\,\beta_1$ = -130

3 **a** β_1 = expected change in flow rate y associated with a one inch increase in pressure drop x = .095

b We expected flow rate to decrease by $5\,\beta_1$ = .475.

c $\mu_{Y \bullet 10} = -.12 + .095(10) = .83$, and $\mu_{Y \bullet 15} = -.12 + .095(15) = 1.305$.

d $P(Y > .835) = P\left(Z > \dfrac{.835 - .830}{.025}\right) = P(Z > .20) = .4207$

$P(Y > .840) = P\left(Z > \dfrac{.840 - .830}{.025}\right) = P(Z > .40) = .3446$

e Let Y_1 and Y_2 denote pressure drops for flow rates of 10 and 11, respectively. Then $\mu_{Y \bullet 11} = .925$, so $Y_1 - Y_2$ has expected value $.830 - .925 = -.095$ and s.d. $[(.025)^2 + (.025)^2]^{1/2} = .035355$. Thus $P(Y_1 > Y_2) = P(Y_1 - Y_2 > 0) = P\left(Z > \dfrac{+.095}{.035355}\right) = P(Z > 2.69) = .0036$.

5 **a** β_1 = expected change for a one degree increase = $-.01$, and $10\,\beta_1 = -.1$ is the expected change for a 10 degree increase.

b $\mu_{Y \bullet 200} = 5.00 - .01(200) = 3$, and $\mu_{Y \bullet 250} = 2.5$.

c The probability that the first observation is between 2.4 and 2.6 is

$P(2.4 \le Y \le 2.6) = P\left(\dfrac{2.4 - 2.5}{.075} \le Z \le \dfrac{2.6 - 2.5}{.075}\right)$

$= P(-1.33 \le Z \le 1.33) = .8164$. The probability that any particular one of the other four observations is between 2.4 and 2.6 is also .8164, so the probability that all five are between 2.4 and 2.6 is $(.8164)^5 = .3627$.

d Let Y_1 and Y_2 denote the times at the higher and lower temperatures, respectively. Then $Y_1 - Y_2$ has expected value $5.00 - .01(x+1) - (5.00 - .01x)$ $= -.01$. The standard deviation of $Y_1 - Y_2$ is $[(.075)^2 + (.075)^2]^{1/2} = .10607$.

Thus $P(Y_1 - Y_2 > 0) = P\left(Z > \dfrac{-(-.01)}{.10607}\right) = P(Z > .09) = .4641$.

Chapter 12

Section 12.2

7 $n = 10, \sum x_i = 1269, \sum y_i = 475, \sum x_i^2 = 172{,}809, \sum y_i^2 = 23{,}251, \sum x_i y_i = 62{,}631$

a $10b_0 + 1269b_1 = 475,\ 1269b_0 + 172{,}809b_1 = 62{,}631$

b $\hat{\beta}_1 = \dfrac{10(62{,}631) - (1269)(475)}{10(172{,}809) - (1269)^2} = \dfrac{23{,}535}{117{,}729} = .19990826$

$\hat{\beta}_0 = \dfrac{475 - .19990826(1269)}{10} = 22.13164131$, so use the equation

$y = 22.13 + .200x$; substituting $\hat{\beta}_0$ and $\hat{\beta}_1$ for b_0 and b_1 in **a** verifies that the normal equations are satisfied.

c estimated expected change = $\hat{\beta}_1 = .200$

d estimated expected change = $12\hat{\beta}_1$ (since 1 yr. = 12 mos.) = 2.4

e 10 yrs. = 120 mos., so $\hat{\mu}_{Y \cdot 120} = \hat{\beta}_0 + \hat{\beta}_1(120) = 46.13$.

9 $n = 14, \Sigma x_i = 3300, \Sigma y_i = 5010, \Sigma x_i^2 = 913{,}750, \Sigma y_i^2 = 2{,}207{,}100, \Sigma x_i y_i = 1{,}413{,}500$

a $\hat{\beta}_1 = \dfrac{3{,}256{,}000}{1{,}902{,}500} = 1.71143233,\ \hat{\beta}_0 = -45.55190543$, so we use the equation

$y = -45.5519 + 1.7114x$.

b $\hat{\mu}_{Y \cdot 225} = -45.5519 + 1.7114(225) = 339.51$

c Estimated expected change $= -50\hat{\beta}_1 = -85.57$

d No, the value 500 is outside the range of x values for which observations were available (the danger of extrapolation).

11 a Verification

b $\hat{\beta}_1 = \dfrac{491.4}{744.16} = .66034186,\ \hat{\beta}_0 = -2.18247148$, so the estimated equation is

$\hat{y} = -2.182 + .660x$.

c predicted $y = \hat{\beta}_0 + \hat{\beta}_1(15) = 7.72$

d $\mu_{Y \cdot 15} = \hat{\beta}_0 + \hat{\beta}_1(15) = 7.72$

Chapter 12

13 **a** $\hat{\beta}_1 = \frac{-404.325}{54,933.75} = -.00736023$, $\hat{\beta}_0 = 1.41122185$,

$SSE = 7.8518 - (1.41122185)(10.68) - (-.00736023)(987.645) = .049245$,

$s^2 = \frac{.049245}{13} = .003788$, and $\hat{\sigma} = s = .06155$

b $SST = 7.8518 - \frac{(10.68)^2}{15} = .24764$ so $r^2 = 1 - \frac{.049245}{.24764} = 1 - .199 = .801$

15 **a**

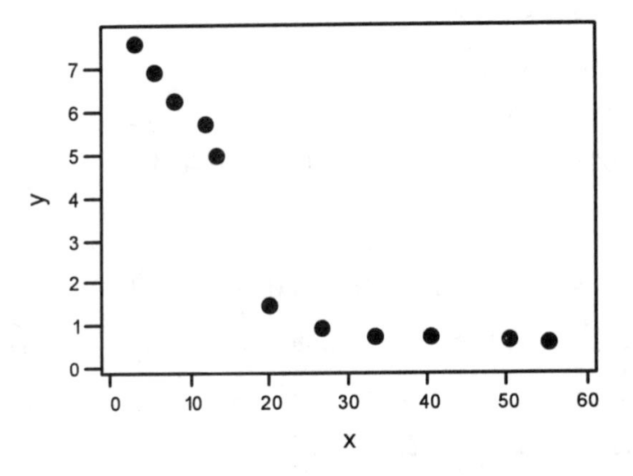

According to a scatter plot of the data, a simple linear regression model does not appear to be plausible. The relationship between the two variables appears to be linear in the range $x \approx 0$ to $x \approx 25$, and at higher values of x the relationship levels off.

b $n = 11$, $\sum x_i = 263.53$, $\sum y_i = 36.66$, $\sum x_i^2 = 9677.4709$, $\sum y_i^2 = 209.7642$, and $\sum x_i y_i = 400.5225$. This gives $\hat{\beta}_1 = \frac{-5255.2623}{37,004.119} = -.14201831$ and $\hat{\beta}_0 = 6.73509866$, so the estimated equation is $y = 6.735 - .14202x$. For $x_8 = 33$, $\hat{y}_8 = 2.048$ and the residual is $.720 - 2.048 = -1.328$.

c $SSE = 19.737$, so $s^2 = \frac{19.737}{9} = 2.193$ and $\hat{\sigma} = s = 1.481$.

d $SST = 87.5864$, so $r^2 = 1 - \frac{19.737}{87.5864} = .775$. The percentage of variation is 77.5%.

Chapter 12

17 Substitution of $\hat{\beta_0} = \dfrac{\sum y_i - \hat{\beta_1}\sum x_i}{n}$ and $\hat{\beta_1}$ for b_0 and b_1 on the lhs of the normal equations yields

$$\frac{n(\sum y_i - \hat{\beta_1}\sum x_i)}{n} + \hat{\beta_1}\sum x_i = \sum y_i \text{ from the first equation and}$$

$$\frac{\sum x_i(\sum y_i - \hat{\beta_1}\sum x_i)}{n} + \hat{\beta_1}\sum x_i^2 = \frac{\sum x_i \sum y_i}{n} + \frac{\hat{\beta_1}(n\sum x_i^2 - (\sum x_i)^2)}{n}$$

$$= \frac{\sum x_i \sum y_i}{n} + \frac{n\sum x_i y_i}{n} - \frac{\sum x_i \sum y_i}{n} = \sum x_i y_i \text{ from the second equation.}$$

19 We wish to find b_1 to minimize $\sum (y_i - b_1 x_i)^2 = f(b_1)$. Equating $f'(b_1)$ to 0 yields $2\sum (y_i - b_1 x_i)(-x_i) = 0$ so $\sum x_i y_i = b_1 \sum x_i^2$ and $b_1 = \dfrac{\sum x_i y_i}{\sum x_i^2}$. The least-squares estimator of β_1 is thus $\hat{\beta_1} = \dfrac{\sum x_i Y_i}{\sum x_i^2}$.

21 For data set #1, $r^2 = .43$ and $\hat{\sigma} = s = 4.03$; whereas these quantities are .99 and 4.03 for #2, and .99 and 1.90 for #3. In general, one hopes for both large r^2 (large % of variation explained) and small s (indicating that observations don't deviate much from the estimated line). Simple linear regression would thus seem to be most effective in the third situation.

Section 12.3

23 **a** $\hat{\beta_1} = -.00736023$, $\hat{\beta_0} = 1.41122185$, so

$SSE = 7.8518 - (1.41122185)(10.68) - (-.00736023)(987.645) = .04925$,

$s^2 = .003788$, $s = .06155$. $\hat{\sigma}^2_{\hat{\beta}_1} = \dfrac{s^2}{\sum x_i^2 - (\sum x_i)^2/n} = \dfrac{.003788}{3662.25} = .00000103$,

$\hat{\sigma}_{\hat{\beta}_1} = s_{\hat{\beta}_1} =$ estimated s.d. of $\hat{\beta_1} = (.00000103)^{1/2} = .001017$.

b $-.00736 \pm (2.160)(.001017) = -.00736 \pm .00220 = (-.00956, -.00516)$

25 **a** H_0: $\beta_1 = 0$

H_a: $\beta_1 \neq 0$

RR: $|t| > t_{\alpha/2, n-2}$ or $|t| > 3.106$

$t = 5.29$

Reject H_0, the slope differs significantly from 0, and the model appears to be useful.

b At the level $\alpha = 0.01$ reject H_0 if the P-value is less than 0.01. In this case reported P-value was 0.000, therefore reject H_0. Conclusion same as for part **a**.

c H_0: $\beta_1 = 1.5$

H_a: $\beta_1 < 1.5$

RR: $t < -t_{\alpha, n-2}$ or $t < -2.718$

$$t = \frac{0.9668 - 1.5}{0.1829} = -2.92$$

Reject H_0. The data contradicts the prior belief.

27 **a** We reject H_0 if $t \geq t_{.01,13} = 2.650$. With $\sum x_i^2 - \frac{(\sum x_i)^2}{n} = 324.40$,

$t = \frac{1.7035 - 1}{3.725/\sqrt{324.40}} = \frac{.7035}{.2068} = 3.40$. Since $3.40 \geq 2.650$, H_0 is rejected in favor of H_a.

b $t_{.005,13} = 3.012$, so the C.I. is $1.7035 \pm \frac{(3.012)(3.725)}{\sqrt{324.40}} = 1.7035 \pm .6229$ $= (1.08, 2.32)$.

29 We wish to test H_0: $\beta_1 = 0$ versus H_a: $\beta_1 \neq 0$. The test statistic value is $t = \frac{\hat{\beta}_1}{s_{\hat{\beta}_1}}$ and H_0 will be rejected at level .05 if either $t \geq t_{.025,9} = 2.262$ or $t \leq -2.262$. With $s = 1.481$ and $\sum (x_i - \bar{x})^2 = \sum x_i^2 - \frac{(\sum x_i)^2}{n} = 3364.0108$, $s_{\hat{\beta}_1} = \frac{1.481}{\sqrt{3364.0108}} = .0255$ and $t = \frac{-.142}{.0255} = -5.6$. Because $-5.6 \leq -2.262$, H_0 is rejected. We conclude that the model does specify a useful relationship.

31 $SSE = 5{,}390{,}382 - (-12.84159)(7034) - (36.18385)(149{,}354.4) = 76{,}492.54$, and $SST = 892{,}458.73$

Source	df	SS	MS	f
Regr	1	815,966.19	815,966.19	96.0
Error	9	76,492.54	8499.17	
Total	10	892,458.73		

Since no α is specified, let's use $\alpha = .01$. Then $F_{.01,1,9} = 10.56 < 96.0$, so H_0: $\beta_1 = 0$ is rejected and the model is judged useful.

$s = \sqrt{8499.17} = 92.19$, $\sum (x_i - \bar{x})^2 = 623.2218$, so $t = \dfrac{36.184}{92.19/\sqrt{623.2218}} = 9.80$ and $t^2 = (9.80)^2 = 96.0 = f$.

33 **a** Let $c = n\sum x_i^2 - (\sum x_i)^2$.

Then $E(\hat{\beta}_1) = \frac{1}{c}E[n\sum x_i Y_i \dots Y_i - (\sum x_i) \dots (\sum x_i)(\sum Y_i)]$

$\frac{n}{c}\sum x_i E(Y_i) - \frac{\sum x_i}{c}\sum E(Y_i) = \frac{n}{c}\sum x_i(\beta_0 + \beta_1 x_i) - \frac{(\sum x_i)}{c}\sum(\beta_0 + \beta_1 x_i)$

$= \frac{\beta_1}{c}[n\sum x_i^2 - (\sum x_i)^2] = \beta_1.$

b With $c = \sum (x_i - \bar{x})^2$, $\hat{\beta}_1 = \frac{1}{c}\sum (x_i - \bar{x})(Y_i - \bar{Y}) = \frac{1}{c}\sum (x_i - \bar{x})Y_i$

(since $\sum (x_i - \bar{x})\bar{Y} = \bar{Y}\sum (x_i - \bar{x}) = \bar{Y}\bullet 0 = 0$), so $V(\hat{\beta}_1) = \frac{1}{c^2}\sum (x_i - \bar{x})^2 \bullet \text{Var}(Y_i)$

$= \frac{1}{c^2}\sum (x_i - \bar{x})^2 \bullet \sigma^2 = \dfrac{\sigma^2}{\sum (x_i - \bar{x})^2} = \dfrac{\sigma^2}{\sum x_i^2 - (\sum x_i)^2/n}$ as desired.

35 The numerator of d is $|1 - 2| = 1$, and the denominator is $\dfrac{4\sqrt{14}}{\sqrt{324.40}} = .831$, so $d = \dfrac{1}{.831} = 1.20$. The appropriate power curve is for $n - 2$ df $= 13$, and β is read from Table A.13 as approximately .1.

Section 12.4

37 **a** $SSE = 45.3623$, $s^2 = 6.4803$, and $s = 2.546$. With $\bar{x} = 2.667$, $n\sum x_i^2 - (\sum x_i)^2 = 58.50$, and $t_{.025,7} = 2.365$, the CI is

$6.45 + (10.6026)(3) \pm (2.365)(2.546)\left[\frac{1}{9} + \frac{9(2.667 - 3)^2}{58.50}\right]^{1/2}$

$= 38.26 \pm (2.365)(2.546)(.3580) = 38.26 \pm 2.16 = (36.10, 40.42)$. The interval is reasonably narrow, indicating a reasonably precise estimate.

b $38.26 \pm (2.365)(2.546)(1.062) = 38.26 \pm 6.39 = (31.87, 44.65)$

c $\bar{x} = 2.667$, so $x = 2.5$ is closer to $\bar{x}$ than is $x = 3.0$, and the intervals for $x = 2.5$ is narrower.

d $x = 6.0$ is well above the range of x values for which data is available, so it would be dangerous to extrapolate the relationship to such an x value. The interval should not be computed.

39 **a** 0.40 is closer to $\bar{x}$

b $(\hat{\beta}_0 + \hat{\beta}_1(0.40)) \pm t_{\alpha/2,\,n-2} \cdot S_{\hat{\beta}_0} + \hat{\beta}_1(0.40)$

$0.8104 \pm (2.101)(0.0311)$

(0.745, 0.876)

c $(\hat{\beta}_0 + \hat{\beta}_1(1.20)) \pm t_{\alpha/2,\,n-2} \cdot \sqrt{s^2 + s^2_{\hat{\beta}_0 + \hat{\beta}_1(1.20)}}$

$0.2912 \pm (2.101) \cdot \sqrt{(0.1049)^2 + (0.0352)^2}$

41 $\hat{\beta}_1 = 18.87349841$, $\hat{\beta}_0 = -8.77862227$, $SSE = 2486.209$, $s = 16.6206$ (431.3, 628.5)

a $\hat{\beta}_0 + \hat{\beta}_1(18) = 330.94$, $\bar{x} = 20.2909$, $\left[\frac{1}{11} + \frac{11(18-20.2909)^2}{3834.26}\right]^{1/2} = .3255$,

$t_{.025,9} = 2.262$, so the C.I. is $330.94 \pm (2.262)(16.6206)(.3255)$ $= 330.94 \pm 12.24 = (318.70, 343.18)$.

b $\left[1 + \frac{1}{11} + \frac{11(18-20.2909)^2}{3834.26}\right]^{1/2} = 1.0516$, so the P.I. is $330.94 \pm (2.262)(16.6206)(1.0516) = 330.94 \pm 39.54 = (291.40, 370.48)$.

c To obtain simultaneous confidence of at least 90% for the two intervals, we compute each one using confidence level 95% (with $t_{.025,9} = 2.262$). For $x = 15$ the interval is $274.32 \pm 15.56 = (258.76, 289.88)$ and for $x = 20$ the interval is $368.69 \pm 11.34 = (357.35, 380.03)$.

43 If each CI is computed using confidence level 98% ($\alpha = .02$), the simultaneous confidence level will be at least $100[1-3(.02)]\% = 94\%$. With $t_{.01,5} = 3.365$, the intervals appear below.

$x = 1200$: $350.3 \pm (3.365)(17.62)(.3936) = 350.3 \pm 23.34$

$x = 1250$: $382.3 \pm (3.365)(17.62)(.3784) = 382.3 \pm 22.44$

$x = 1300$: $414.3 \pm (3.365)(17.62)(.4054) = 414.3 \pm 24.03$

45 A 95% prediction interval for a future y to be observed when $x = 15$ is

$$6.74 + (-.1420)(15) \pm (2.262)(1.481)\left[1 + \frac{1}{11} + \frac{11(15-23.957)^2}{37{,}004.119}\right]^{1/2}$$

$$= 4.61 \pm (2.262)(1.481)(1.056) = 4.61 \pm 3.54 = (1.07,\ 8.15)$$

The prediction interval is rather wide.

Chapter 12

47 **a** $x_2 = x_3 = 12$ yet $y_2 \neq y_3$.

b

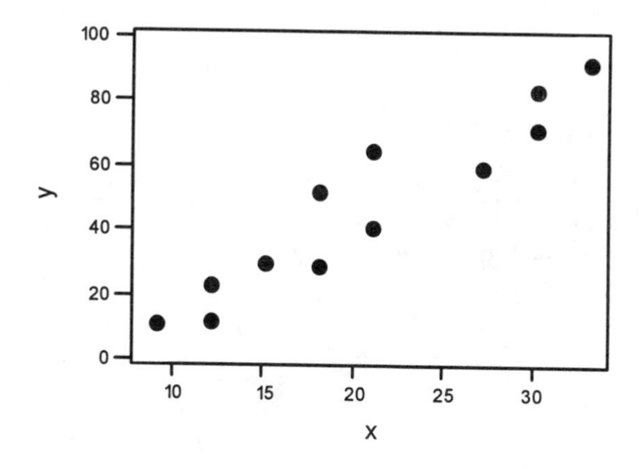

Based on a scatter plot of the data, a simple linear regression model does seem a reasonable way to describe the relationship between the two variables.

c $\hat{\beta}_1 = \dfrac{2296}{699} = 3.284692,\ \hat{\beta}_0 = -19.669528,\ y = -19.67 + 3.2847x$

d $SSE = 35{,}634 - (-19.669528)(572) - (3.284692)(14{,}022)$

$= 827.0188,\ s^2 = 82.70188,\ s = 9.094$

$$s_{\hat{\beta}_0 + \hat{\beta}_1(20)} = 9.094\sqrt{\frac{1}{12} + \frac{12(20.5 - 20)^2}{8388}} = 2.6308$$

$\hat{\beta}_0 + \hat{\beta}_1(20) = 46.03,\ t_{.025,10} = 2.228$. The P.I. is

$$46.03 \pm 2.228\sqrt{s^2 + s^2_{\hat{\beta}_0 + \hat{\beta}_1(20)}} = 46.03 \pm 21.09 = (24.94, 67.12).$$

Section 12.5

49 **a** $n = 10,\ \sum x_i = 213.80,\ \sum y_i = 4835,\ \sum x_i^2 = 5158.1346,\ \sum y_i^2 = 2{,}659{,}583,$ and $\sum x_i y_i = 113{,}582.77$, so

$$r = \frac{10(113{,}582.77) - (213.80)(4835)}{[10(5158.1346) - (213.80)^2]^{1/2}[10(2{,}659{,}583) - (4835)^2]^{1/2}}$$

$$= \frac{102{,}104.70}{(76.622)(1794.047)} = .7428.$$

b $r^2 = (.7428)^2 = .552$, so 55.2% of variation in y is explained by the model.

Chapter 12

51 **a** $r = .9066$, $t = 7.75$, H_0: $\rho = 0$ versus H_a: $\rho > 0$, and the P-value for an upper-tailed test satisfies P-value < .0005, so H_0 would be rejected at any reasonable significance level.

b $r^2 = (.9066)^2 = .822$

53 $n = 6$, $\Sigma x = 111.71$, $\Sigma x^2 = 2{,}724.7943$

$\Sigma y = 2.9$, $\Sigma y^2 = 1.6572$, $\Sigma xy = 63.915$

$$r = \frac{(6)(63.915) - (111.71)(2.9)}{\sqrt{(6)(2{,}724.7943 - (111.71)^2} \cdot \sqrt{(6)(1.6572) - (2.9)^2}}$$

$r = .7729$

H_0: $\rho = 0$

H_a: $\rho \neq 0$

$R.R.$ $|t| \geq t_{\alpha/2,4}$ or $|t| \geq 2.776$

$$t = \frac{(.7729)\sqrt{4}}{\sqrt{1 - (.7729)^2}} = 2.436$$

Fail to reject H_0. The data does not indicate that the population correlation coefficient differs from 0. This result may seem surprising due to the relatively large size of r (.77), however, it can be attributed to a small sample size (6).

55 **a** Although the normal probability plot of the x's appears somewhat curved, such a pattern is not terribly unusual when n is small; the test of normality presented in Section 14.2 (p. 604) does not reject the hypothesis of population normality. The normal probability plot of the y's is much straighter.

b H_0: $\rho = 0$ will be rejected in favor of H_a: $\rho \neq 0$ if either $t \geq t_{.005,8} = 3.355$ or $t \leq -3.355$. $\sum x_i = 864$, $\sum y_i = 138.0$, $\sum x_i^2 = 78{,}142$, $\sum x_i y_i = 12{,}322.4$, $\sum y_i^2 = 1959.10$, so $r = \dfrac{3992}{(186.8796)(23.3880)} = .913$ and

$t = \dfrac{.913(2.8284)}{.4080} = 6.33$. Because $6.33 \geq 3.355$, H_0 is rejected. There does appear to be a linear relationship.

57 $t = 2.20 \geq t_{.025,9998} = 1.96$, so H_0 is rejected in favor of H_a. The value $t = 2.20$ is statistically significant — it cannot be attributed just to sampling variability in the case $\rho = 0$. But with this n, $r = .022$ implies $\rho = 0.22$, which in turn shows an extremely weak linear relationship.

Chapter 12

Supplementary

59 **a** The test statistic value is $t = \frac{\hat{\beta}_1 - 1}{s_{\hat{\beta}_1}}$, and H_0 will be rejected if either $t \geq t_{.025,11} = 2.201$ or $t \leq -2.201$. With $\sum x_i = 243$, $\sum y_i = 241$, $\sum x_i^2 = 5965$, $\sum y_i^2 = 5731$, and $\sum x_i y_i = 5805$, $\hat{\beta}_1 = .913819$, $\hat{\beta}_0 = 1.457072$, $SSE = 75.126$, $s = 2.613$, and $s_{\hat{\beta}_1} = .0693$, $t = \frac{.9138 - 1}{.0693} = -1.24$. Because -1.24 is neither ≤ -2.201 nor ≥ 2.201, H_0 cannot be rejected. It is plausible that $\beta_1 = 1$.

b $r = \frac{16{,}902}{(136)(128.15)} = .970$

61 **a** $n = 16$

$$r = \frac{(16)(79{,}574) - (8{,}140)(145.7)}{\sqrt{(16)(4{,}340{,}600) - (8{,}140)^2} \cdot \sqrt{(16)(1{,}505.01) - (145.7)^2}}$$

$r = .914$

$r^2 = .836$

Thus about 84% of the variation in shoot elongation is explained by variation in soil heat.

b H_0: $\rho = 0$

H_a: $\rho \neq 0$

RR: $|t| \geq t_{0.025,14}$ or $|t| \geq 2.145$

$$t = \frac{(.914)\sqrt{14}}{\sqrt{1 - (.914)^2}} = 8.43$$

Reject H_0, there does appear to be a useful linear relationship

c First compute the regression line

$$\hat{\beta}_1 = \frac{(16)(79{,}574) - (8140)(145.7)}{(16)(4{,}340{,}600) - (8140)^2} = .02733$$

$$\hat{\beta}_0 = \left(\frac{145.7}{16}\right) - (0.02733)\left(\frac{8140}{16}\right) = -4.7979$$

$SSE = (1505.01) - (-4.7979)(145.7) - (.02733)(79{,}574) = 29.30661$

$$s^2 = \frac{29.30661}{14} = 2.0933$$

$s = 1.4468$

when $x = 500$, $\hat{y} = 8.8671$

$$s_{\hat{\beta}_0+\hat{\beta}_1 x^*} = (1.4468)\sqrt{\frac{1}{16}+\frac{16(500-508.75)^2}{16(4,340,600)-(8140)^2}} = .3628$$

H_0: $\mu_{y\cdot 500} = 10$

H_a: $\mu_{y\cdot 500} \neq 10$

RR: $|t| \geq t_{0.025,14}$ or $|t| \geq 2.145$

$$t = \frac{8.8671-10}{0.3628} = -3.12$$

Reject H_0. The data indicates that the expected amount of shoot elongation differs significantly from 10 when daily degree hours of soil heat equals 500.

63 **a** $n = 9$, $\sum x_i = 228$, $\sum y_i = 93.76$, $\sum x_i^2 = 5958$, $\sum y_i^2 = 982.2932$, and $\sum x_i y_i = 2348.15$, giving $\hat{\beta}_1 = \frac{-243.93}{1638} = -.148919$, $\hat{\beta}_0 = 14.190392$, and the equation $y = 14.19 - .1489x$.

b β_1 is the expected increase in load associated with a one-day age increase (so a negative value of β_1 corresponds to a decrease). We wish to test H_0: $\beta_1 = -.10$ versus H_a: $\beta_1 < -.10$ (the alternative contradicts prior belief). H_o will be rejected at level .05 if $t = \frac{\hat{\beta}_1-(-.10)}{s_{\hat{\beta}_1}} \leq -t_{.05,7} = -1.895$. With SSE $= 1.4862$, $s = .4608$ and $s_{\hat{\beta}_1} = \frac{.4608}{\sqrt{182}} = .0342$. Thus $t = \frac{-.1489+.1}{.0342} = -1.43$. Because -1.43 is not ≤ -1.895, don't reject H_0.

c $\sum x_i = 306$ and $\sum x_i^2 = 7946$, so $\sum(x_i-\bar{x})^2 = 7946 - \frac{(306)^2}{12} = 143$ here, as contrasted with 182 for the given 9 x_i's. Even though the sample size for the proposed x values is larger, the original set of values is preferable.

d $\left[\frac{1}{9}+\frac{9(28-25.33)^2}{1638}\right]^{1/2} = .3877$ so $(t_{.025,7})(s)[\ ]^{1/2} = (2.365)(.4608)(.3877) = .42$, and $\hat{\beta}_0 + \hat{\beta}_1(28) = 10.02$, so the 95% C.I. is $10.02 \pm .42 = (9.60, 10.44)$.

65 **a** The plot suggests a strong linear relationship between x and y.

b $n = 9$, $\sum x_i = 179.7$, $\sum y_i = 7.28$, $\sum x_i^2 = 4334.41$, $\sum y_i^2 = 7.4028$, and $\sum x_i y_i = 178.683$, so $\hat{\beta}_1 = \frac{299.931}{6717.6} = .04464854$, $\hat{\beta}_0 = -.08259353$, and the equation of the estimated line is $y = -.08259 + .044649x$.

c $SSE = 7.4028-(-.601281)-7.977935 = .026146$,

$SST = 7.4028-\frac{(7.28)^2}{9} = 1.5141$, and $r^2 = 1-\frac{SSE}{SST} = .983$, so 98.3% of the observed variation is "explained".

d $\hat{y}_4 = -.08259+(.044649)(19.1) = .7702$, and $y_4-\hat{y}_4 = .68-.7702 = -.0902$

e $s = .06112$ and $s_{\hat{\beta}_1} = \frac{.06112}{\sqrt{746.4}} = .002237$, so the value of t for testing

H_o: $\beta_1 = 0$ versus H_a: $\beta_1 \neq 0$ is $\frac{.044649}{.002237} = 19.96$. From Table A.5, $t_{.0005,7} = 5.408$, so P-value $< 2(.0005) = .001$. There is strong evidence for a useful relationship.

f A 95% CI for β_1 is $.044649 \pm (2.365)(.002237) = .044649 \pm .005291$ $= (.0394,.0499)$.

g A 95% CI for $\beta_0+\beta_1(20)$ is $.810 \pm (2.365)(.06112)(.3333356) = .810 \pm .048$ $= (.762,.858)$.

67 a With $s_{xx} = \sum(x_i-\bar{x})^2$, $s_{yy} = \sum(y_i-\bar{y})^2$, note that $\frac{s_y}{s_x} = \sqrt{\frac{s_{yy}}{s_{xx}}}$ (since the factor $n-1$ appears in both the numerator and denominator, so cancels). Thus

$$y = \hat{\beta}_0+\hat{\beta}_1 x = \bar{y}+\hat{\beta}_1(x-\bar{x}) = \bar{y}+\frac{s_{xy}}{s_{xx}}(x-\bar{x}) = \bar{y}+\sqrt{\frac{s_{yy}}{s_{xx}}}\bullet\frac{s_{xy}}{\sqrt{s_{xx}s_{yy}}}(x-\bar{x})$$

$$= \bar{y}+\frac{s_y}{s_x}\bullet r\bullet(x-\bar{x}) \text{ as desired.}$$

b By .573 sd's above (above since $r < 0$), or (since $s_y = 4.3143$) an amount 2.4721 above.

69 Using the notation of the exercise above, $SST = s_{yy}$ and $SSE = s_{yy}-\hat{\beta}_1 s_{xy}$ $= s_{yy}-\frac{s_{xy}^2}{s_{xx}}$, so $1-\frac{SSE}{SST} = 1-\frac{s_{yy}-s_{xy}^2/s_{xx}}{s_{yy}} = \frac{s_{xy}^2}{s_{xx}s_{yy}} = r^2$ as desired.

71 For the second boiler, $n = 6$, $\sum x_i = 125$, $\sum y_i = 472.0$, $\sum x_i^2 = 3625$, $\sum y_i^2 = 37{,}140.82$, and $\sum x_i y_i = 9749.5$, giving $\hat{\gamma}_1 =$ estimated slope $= \frac{-503}{6125} = -.0821224$, $\hat{\gamma}_0 = 80.377551$, $SSE_2 = 3.26827$, and $SSx_2 = 1020.833$. For boiler #1, $n = 8$, $\hat{\beta}_1 = -.1333$, $SSE_1 = 8.733$, and $SSx_1 = 1442.875$. Thus

$\hat{\sigma}^2 = \dfrac{8.733 + 3.268}{10} = 1.2$, $\hat{\sigma} = 1.095$, and $t = \dfrac{-.1333 + .0821}{1.095(1/1442.875 + 1/1020.833)^{1/2}}$ $= \dfrac{-.0512}{.0448} = -1.14$. $t_{.025,10} = 2.228$ and -1.14 is neither ≥ 2.228 nor ≤ -2.228, so H_0 is not rejected. It is plausible that $\beta_1 = \gamma_1$.

CHAPTER 13

Section 13.1

1 **a** $\bar{x} = 15$ and $\Sigma(x_j - \bar{x})^2 = 250$, so the s.d. of $Y_i - \hat{Y}_i$ is $10[1 - 1/5 - (x_i - 15)^2/250]^{1/2}$ = 6.32, 8.37, 8.94, 8.37, and 6.32 for i = 1, 2, 3, 4, 5.

b Now $\bar{x} = 20$ and $\Sigma(x_j - \bar{x})^2 = 1250$, giving standard deviations 7.87, 8.49, 8.83, 8.94, and 2.83 for i = 1, 2, 3, 4, 5.

c The deviation from the estimated line is likely to be much smaller for the observation made in the experiment of (b) for $x = 50$ than for the experiment of (a) when $x = 25$. That is, the observation (50, Y) is more likely to fall close to the least-squares line than is (25, Y).

3 **a** This plot indicates there are no outliers, the variance of ϵ is constant, and the ϵ are normally distributed. A straight-line regression function is a reasonable choice for a model.

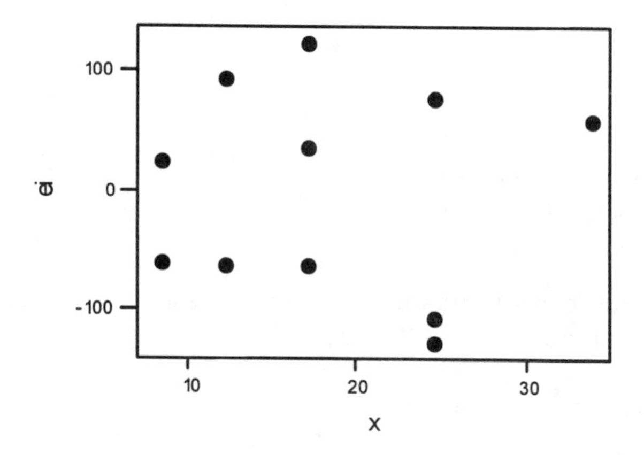

b With $\Sigma(x_j - \bar{x})^2 = \Sigma x_j^2 - \dfrac{(\Sigma x_j)^2}{n} = 4198.03 - 3574.81 = 623.22$

$e_i^* = \dfrac{e_i}{92.19}\left[1 - \dfrac{1}{11} - \dfrac{(x_i - 18.03)^2}{623.22}\right]^{-1/2}$. This gives e_i^* = -.75, .31, -.74, 1.13, .43, -.72, 1.43, .93, -1.51, -1.27, .90 for i = 1,...,11; $\dfrac{e_2}{s} = \dfrac{24.52}{92.19} = .27$ while $e_2^* = .31$, and $\dfrac{e_7}{s} = \dfrac{125.72}{92.19} = 1.36 \neq 1.43 = e_7^*$, so $\dfrac{e_i}{s}$ is not $\doteq e_i^*$.

c The plot of e_i^* vs. x has almost exactly the same pattern as the plot of (a); only the scale on the vertical axis is changed.

5 The standardized residuals (in order corresponding to increasing x) are -.50, -.75, -.50, .79, .90, .93, .19, 1.46, -1.80, and -1.12. A standardized residual plot shows the same pattern as the residual plot discussed in the previous exercise.

7 **a**

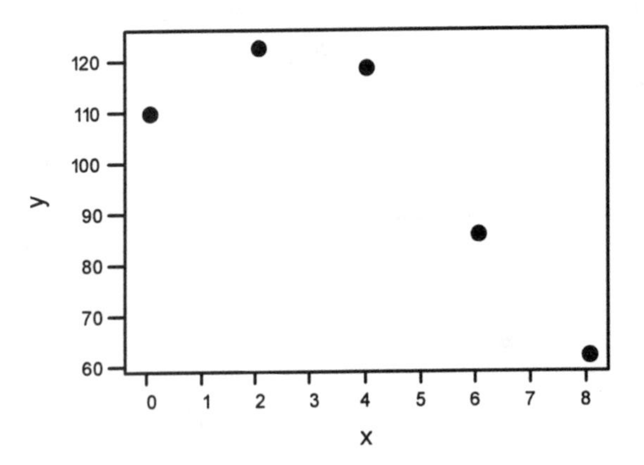

There is an obvious curved pattern in the scatter plot, which suggests that a simple linear model will not provide a good fit.

b The $\hat{y}$'s, e's, and e^*'s are given below.

x	y	$\hat{y}$	e	e^*
0	110	126.6	-16.6	-1.55
2	123	113.3	9.7	.68
4	119	100.0	19.0	1.25
6	86	86.7	-.7	-.05
8	62	73.4	-11.4	-1.06

9 Both a scatter plot and residual plot (based on the simple linear regression model) for the first data set suggest that a simple linear regression model is reasonable, with no pattern or influential data points which would indicate that the model should be modified. However, scatter plots for the other three data sets reveal "difficulties."

Chapter 13

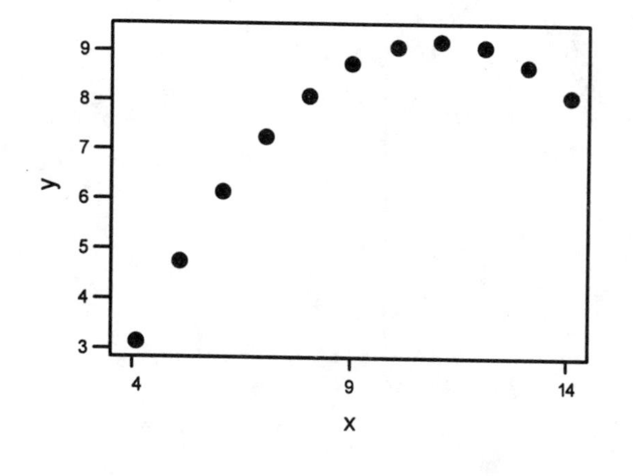

#2

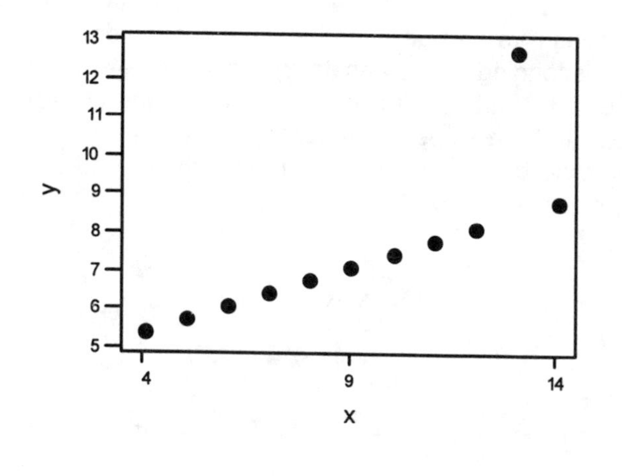

#3

Chapter 13

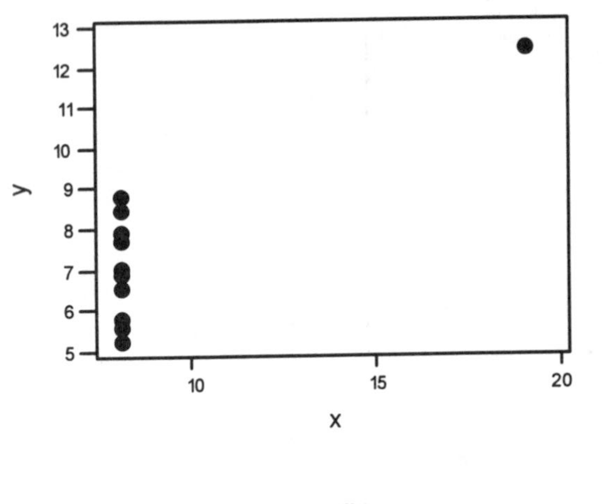

#4

For data set #2, a quadratic function would clearly provide a much better fit. For data set #3, the relationship is perfectly linear except for the outlier, which has obviously greatly influenced the fit even though its x value is not unusually large or small; the signs of the residuals here (corresponding to increasing x) are ++++-----+-, and a residual plot would reflect this pattern and suggest a careful look at the chosen model. For data set #4 it is clear that the slope of the least squares line has been determined entirely by the outlier, so this point is extremely influential (and its x value does lie far from the remaining ones).

11 **a** $Y_i - \hat{Y}_i = Y_i - \bar{Y} - \hat{\beta}_1(x_i - \bar{x}) = Y_i - \frac{1}{n}\sum_j Y_j - \frac{(x_i - \bar{x})\sum_j (x_j - \bar{x})Y_j}{\sum_j (x_j - \bar{x})^2} = \sum_j c_j Y_j$ where

$c_j = 1 - \frac{1}{n} - \frac{(x_i - \bar{x})^2}{n\sum(x_j - \bar{x})^2}$ for $j = 1$ and $c_j = -\frac{1}{n} - \frac{(x_i - \bar{x})(x_j - \bar{x})}{\sum(x_j - \bar{x})^2}$ for $j \neq i$. Thus

$Var(Y_i - \hat{Y}_i) = \sum Var(c_j Y_j)$ (since the Y_j's are independent) $= \sigma^2 \sum c_j^2$ which, after some algebra, gives (13.2).

b $\sigma^2 = Var(Y_i) = Var(\hat{Y}_i + (Y_i - \hat{Y}_i)) = Var(\hat{Y}_i) + Var(Y_i - \hat{Y}_i)$, so

$Var(Y_i - \hat{Y}_i) = \sigma^2 - Var(\hat{Y}_i) = \sigma^2 - \sigma^2\left[\frac{1}{n} + \frac{(x_i - \bar{x})^2}{\sum(x_j - \bar{x})^2}\right]$, which is exactly (13.2).

c As x_i moves further from $\bar{x}$, $(x_i - \bar{x})^2$ grows larger, so $Var(\hat{Y}_i)$ increases (since $(x_i - \bar{x})^2$ has a positive sign in $Var(\hat{Y}_i)$) but $Var(Y_i - \hat{Y}_i)$ decreases (since $(x_i - \bar{x})^2$ has a negative sign).

13 The distribution of any particular standardized residual is also a t distribution with $n-2$ df, since e_i^* is obtained by taking the standard normal variable $(Y_i - \hat{Y}_i)/(sd \text{ of } Y_i - \hat{Y}_i)$ and substituting the estimate of σ in the denominator (exactly as in the predicted value case). With E_i^* denoting the ith standardized residual as a rv, when $n = 25$ E_i^* has a t distribution with 23 df and $t_{.01,23} = 2.50$ so $P(E_i^*$ outside $(-2.50, 2.50)) = P(E_i^* \geq 2.50) + P(E_i^* \leq -2.50) = .01 + .01 = .02$.

15 a The summary quantities are (via computer) $\Sigma x_i' = 40.296$, $\Sigma y_i' = 30.330$, $\Sigma x_i'^2 = 148.91$, $\Sigma y_i'^2 = 89.801$, and $\Sigma x_i' y_i' = 108.31$, from which $\hat{\beta}_1 = -2.16$ and $\hat{\beta}_0 = 10.67$. The model is $Y = \alpha x^{\beta} \varepsilon$ where

$\beta = \beta_1$ and $\beta_0 = \ln(\alpha)$, so $\hat{\beta} = -2.16$ and $\hat{\alpha} = e^{\hat{\beta}_0} = 43{,}044.94$.

b Yes (the x_i''s are 3.11, 3.22, 3.33, 3.42, 3.64, 3.70, 3.75, 3.87, 4.00, 4.01, 4.25 with corresponding standardized residuals -1.64, .20, .36, .40, .68, -.57, 1.21, .22, 1.51, -1.61, -1.12).

c Yes. This is not surprising since $r^2 = .979$ for a straight-line fit to the transformed data.

d We wish to test H_0: $\beta = -2$ vs. H_a: $\beta \neq -2$, or equivalently H_0: $\beta_1 = -2$ vs. H_a: $\beta_1 \neq -2$ for the linear-model fit to the transformed data.

The estimated standard deviation of $\hat{\beta}_1$ is $\dfrac{s}{[\Sigma x_i'^2 - (\Sigma x_i')^2/n]^{1/2}} = .1046$, so

$t = \dfrac{-2.16 - (-2)}{.1046} = -1.53$. H_0 will be rejected at level .05 if either $t \geq t_{.025,9} = 2.262$ or if $t \leq -2.262$. Since -1.53 is neither ≥ 2.262 nor ≤ -2.262, H_0 cannot be rejected.

e The CI for β_1 $(= \beta)$ is $-2.16 \pm (2.262)(.1046) = -2.16 \pm .24 = (-2.40, -1.92)$.

f We first obtain a CI for $\beta_0 + \beta_1 x' = \beta_0 + \beta_1(3.6889)$ using the transformed data and then "exponentiate" each endpoint to obtain the desired C.I. With

$\hat{\beta}_0 + \hat{\beta}_1 x' = 2.7018$, $s = .1191$, and $\left[\dfrac{1}{n} + \dfrac{n(x' - \bar{x}')^2}{n\Sigma x'^2 - (\Sigma x')^2}\right]^{1/2} = .302351$, the CI for $\beta_0 + \beta_1 x'$ is $2.7018 \pm (2.262)(.1191)(.302351) = 2.7018 \pm .0815$ = (2.6203, 2.7833). The desired CI is then $(e^{2.6203}, e^{2.7833}) = (13.74, 16.17)$.

17 a $\Sigma x_i' = 15.501$, $\Sigma y_i' = 13.352$, $\Sigma x_i'^2 = 20.228$, $\Sigma y_i'^2 = 16.572$, $\Sigma x_i' y_i' = 18.109$, from which $\hat{\beta}_1 = 1.254$, $\hat{\beta}_0 = -.468$, so $\hat{\beta} = 1.254$, $\hat{\alpha} = e^{-.468} = .626$.

b The plots give strong support to this choice of model; in addition, $r^2 = .960$ for the transformed data.

c $SSE = .11536$ (computer output), $s = .1024$, and the estimated sd of $\hat{\beta}_1$ is .0775, so $t = \dfrac{1.25 - 1.333}{.0775} = -1.07$. SInce -1.07 is not $\le -t_{.05,11} = -1.796$, H_0 cannot be rejected in favor of H_a.

d The claim that $\mu_{Y\cdot5} = 2\mu_{Y\cdot2.5}$ is equivalent to $\alpha \cdot 5^\beta = 2\alpha(2.5)^\beta$, or that $\beta = 1$. Thus we wish to test H_0: $\beta_1 = 1$ vs. H_a: $\beta \ne 1$. With $t = \dfrac{1 - 1.333}{.0775} = -4.30$ and $t_{.005,11} \le 3.106$, H_0 is rejected at level .01 since $-4.30 \le -3.106$.

19 a No. There is definite curvature in the plot.

b $Y' = \beta_0 + \beta_1 x' + \varepsilon$ where $x' = \dfrac{1}{\text{temp.}}$ and $y' = \ln(\text{lifetime})$. Plotting y' vs. x' gives a plot which has a pronounced linear appearance (and in fact $r^2 = .954$ for the straight-line fit).

c $\Sigma x_i' = .082273$, $\Sigma y_i' = 123.64$, $\Sigma x_i'^2 = .00037813$, $\Sigma y_i'^2 = 879.88$, $\Sigma x_i' y_i' = .57295$, from which $\hat{\beta}_1 = 3735.4485$ and $\hat{\beta}_0 = -10.2045$ (values read from computer output). With $x = 220$, $x' = .00445$ so $\hat{y}' = -10.2045 + 3735.4485(.00445) = 6.7748$ and thus $\hat{y} = e^{\hat{y}'} = 875.50$.

d For the transformed data, $SSE = 1.39857$, and $n_1 = n_2 = n_3 = 6$, $\bar{y}_{1.}' = 8.44695$, $\bar{y}_{2.}' = 6.83157$, $\bar{y}_{3.}' = 5.32891$, from which $SSPE = 1.36594$, $SSLF = .02993$, $f = \dfrac{.02993/1}{1.36594/15} = .33$. Comparing this to $F_{.01,1,15} = 8.68$, it is clear that H_0 cannot be rejected.

21 a The suggested model is $Y = \beta_0 + \beta_1 x' + \varepsilon$ where $x' = \dfrac{10^4}{x}$. The summary quantities are $\Sigma x_i' = 159.01$, $\Sigma y_i = 121.50$, $\Sigma x_i'^2 = 4058.8$,

$\Sigma y_i^2 = 1865.2$, $\Sigma x_i' y_i = 2281.6$, from which $\hat{\beta}_1 = -.1485$ and $\hat{\beta}_0 = 18.1391$, and the estimated regression function is $y = 18.1391 - \frac{1485}{x}$.

b $x = 500 \Rightarrow \hat{y} = 18.1391 - \frac{1485}{500} = 15.17.$

23 $Var(Y) = Var(\alpha e^{\beta x} \cdot \varepsilon) = [\alpha e^{\beta x}]^2 \cdot Var(\varepsilon) = \alpha^2 e^{2\beta x} \cdot \tau^2$ where we have set $Var(\varepsilon) = \tau^2$. If $\beta > 0$, this is an increasing function of x so we expect more spread in y for large x than for small x, while the situation is reversed if $\beta < 0$. It is important to realize that a scatter plot of data generated from this model will not spread out uniformly about the exponential regression function throughout the range of x values; the spread will only be uniform on the transformed scale. Similar results hold for the multiplicative power model.

Section 13.3

25

a $\hat{\mu}_{y \cdot 75} = \hat{\beta}_0 + \hat{\beta}_1(75) + \hat{\beta}_2(75)^2 = -113.0937 + 3.3684(75) - .01780(75)^2 = 39.41.$

b $\hat{y} = \hat{\beta}_0 + \hat{\beta}_1(60) + \hat{\beta}_2(60)^2 = 24.93.$

c $SSE = \Sigma y_i^2 - \hat{\beta}_0 \Sigma y_i - \hat{\beta}_1 \Sigma x_i y_i - \hat{\beta}_2 \Sigma x_i^2 y_i = 8386.43 - (-113.0937)(210.70)$ $-(3.3684)(17{,}002) - (-.0178)(1{,}419{,}780) = 217.82$, $s^2 = \frac{SSE}{n-3} = \frac{217.82}{3} = 72.61$, $s = 8.52$

d $R^2 = 1 - \frac{217.82}{987.35} = .779.$

e H_0 will be rejected in favor of H_a if either $t \geq t_{.005,3} = 5.841$ or if $t \leq -5.841$. The computed value of T is $t = \frac{-.01780}{.00226} = -7.88$, and since $-7.88 \leq -5.841$, we reject H_0.

27

a $R^2 = 0.853$. This means 85.3% of the variation in wheat yield is accounted for by the model.

b $(-135.44) \pm 2.201(41.97)$
$(-227.82, -43.06)$

c H_0: $\mu_{y \cdot 2.5} = 1500$
H_a: $\mu_{y \cdot 2.5} < 1500$
RR: $t < -t_{0.01,11}$ or $t < -2.718$
When $x = 2.5$, $\hat{y} = 1402.15$

$t = \dfrac{1402.15 - 1500}{53.5} = -1.83$

Fail to reject H_0. The data does not indicate $\mu_{y \cdot 2.5}$ is less than 1500.

d $1{,}402.15 \pm (2.201)\sqrt{(135.6)^2 + (53.5)^2}$
(1081.37, 1722.99)

29 **a** $.3463 - 1.2933(x-\bar{x}) + 2.3964(x-\bar{x})^2 - 2.3968(x-\bar{x})^3$

b From (a) the coefficient of x^3 is clearly -2.3968, so $\hat{\beta}_3 = -2.3968$. There will be a contribution to x^2 both from $2.3964(x-4.3456)^2$ and from $-2.3968(x-4.3456)^3$. Expanding these and adding yields 33.6430 as the coefficient of x^2, so $\hat{\beta}_2 = 33.6430$.

c $x = 4.5 \Rightarrow x' = x - \bar{x} = .1544$; substituting into (a) yields $\hat{y} = .1949$.

d $t = \dfrac{-2.3968}{2.4590} = -.97$ which is not significant (H_0: $\beta_3 = 0$ cannot be rejected), so the inclusion of the cubic term is not justified.

31 **a** $\bar{x} = 49.9231$, $s_x = 41.3652$, so for $x = 50$, $x' = \dfrac{x - 49.9231}{41.3652} = .001859$ and

$\hat{\mu}_{Y \cdot 50} = .8733 - .3255x' + .0448(x')^2 = .873$.

b $SST = 1.456923$ and $SSE = .117521$, so $R^2 = .919$.

c $.8733 - .3255\dfrac{x - 49.9231}{41.3652} + .0448\dfrac{(x-49.9231)^2}{(41.3652)^2} =$

$1.200887 - .01048314x + .00002618x^2$

d $\hat{\beta}_2 = \dfrac{\hat{\beta}_2^*}{s_x^2}$ so the estimated sd of $\hat{\beta}_2$ is the estimated sd of $\hat{\beta}_2^*$ multiplied by

$\dfrac{1}{s_x}$: $s_{\hat{\beta}_2} = (.0319)\left(\dfrac{1}{41.3652}\right) = .00077118$.

e $t = \dfrac{.0448}{.0319} = 1.40$ which is not significant (compare to $\pm\, t_{.025,9}$ at level .05), so the quadratic term should not be retained.

Section 13.4

33 **a** $\mu_{Y \cdot 6000,8,5} = -70 + .025(6000) + 20(8.0) + 7.5(5.0) = 277.5$

b Expected change $= \beta_2 = \$20$

c In this case, Y has a normal distribution with mean value 277.5 and standard deviation 20, so $P(240 \le Y \le 300) = P\left(\frac{240-277.5}{20} \le Z \le \frac{300-277.5}{20}\right)$
$= P(-1.88 \le Z \le 1.13) = \Phi(1.13) - \Phi(-1.88) = .8708 - .0301 = .8407$

35 **a** $R^2 = 1 - \frac{\text{SSE}}{\text{SST}} = 1 - \frac{6557.7}{21005.9} = 0.6878$
This indicates the model accounts for about 69% of the variation in ball bond shear strength.

b $F_{.01,\,4,\,295}$ is not tabled so use $F_{.01,\,4,\,120} = 3.48$. The computed F statistic is 162.71, therefore reject H_0 and conclude that the model is indeed useful.

c $1{,}236.9687 \pm (1.96)(0.950) = (1235.11, 1238.83)$

d $1.8542 \pm (1.96)(0.3045) = (1.26, 2.45)$

37 $R^2 = 1 - \frac{20.0}{39.2} = .490$. For testing H_0: $\beta_1 = \beta_2 = \beta_3 = \beta_4 = 0$ vs. H_a: at least one among $\beta_1,\ldots,\beta_4$ is not zero, the test statistic is $F = \left(\frac{R^2}{k}\right) / \left[\frac{1-R^2}{n-k-1}\right]$. H_0 will be rejected if $f \ge F_{.05,4,25} = 2.76$. $f = \left(\frac{.490}{4}\right) / \left(\frac{.510}{25}\right) = 6.0$ Because $6.0 \ge 2.76$, H_0 is rejected and the model is judged useful (this even though the value of R^2 is not all that impressive).

39 **a** The test of model utility is appropriate here, with $n = 180$, $k = 4$, and the test statistic $f = \left(\frac{R^2}{4}\right) / \left[\frac{1-R^2}{175}\right]$. H_0 will be rejected in favor of H_a at level .01 if $f \ge F_{.01,4,175} \approx 3.85$. With $R^2 = .820$, the computed value of f is $\left(\frac{.820}{4}\right) / \left(\frac{.180}{175}\right) = 199.3$. Since $199.3 \ge 3.85$, H_0 is (resoundingly) rejected in favor of H_a: y appears to be linearly related to at least one of the four carriers.

b We compute a 98% CI for each of the two β's separately, giving simultaneous confidence of at least 96%. With $t_{.01,175} \doteq z_{.01} = 2.33$, the CI for β_3 is $-.02059 \pm (2.33)(.0027) = (-.02688, -.01430)$ and the CI for β_4 is $-.10155 \pm (2.33)(.0373) = (-.18846, -.01464)$.

c $s^2 = \frac{SSE}{n-(k+1)} = \frac{1275.75}{175} = 7.290$, $s = 2.700$. $R^2 = .820 =$
$1 - \frac{SSE}{SST} \Rightarrow \frac{SSE}{SST} = .180 \Rightarrow SST = \frac{1275.75}{.180} = 7087.50.$

d The appropriate test procedure is given by (13.24) with $k = 6$ (for the full model), $SSE_k = 1247.30$, $l = 4$ (for the reduced model which contains carriers

x_1, x_2, x_1^2, x_2^2), and SSE_l = 1275.75. H_0 will be rejected if $f \geq F_{.05,2,173} \doteq 3.0$.

The computed value of f is $f = \left(\frac{1275.75 - 1247.30}{2}\right) / \left(\frac{1247.30}{173}\right) =$

$\frac{14.225}{7.210} = 1.97$. Since 1.97 is not ≥ 3.0, H_0 cannot be rejected. There is no reason to include either x_1^3 or x_2^3 as carriers in the model.

41 $SSE_k = 23.379$, $SSE_l = 203.82$

H_0: $\beta_4 = \beta_5 = \ldots = \beta_9 = 0$

H_a: at least one of the above $\beta_i \neq 0$

RR: $f \geq f_{\alpha, k-l, n-(k+1)}$ or $f \geq F_{0.05,6,5}$ or $f \geq 4.95$

$$f = \frac{(203.82 - 23.379)/(9-3)}{23.379/5} = 6.43$$

Reject H_0. At least one of the second-order predictors appears useful.

43

a Here $k = 5$, $n-(k+1) = 6$, so H_0 will be rejected in favor H_a at level .05 if either $t \geq t_{.025,6} = 2.447$ or $t \leq -2.447$. The computed value of T is $t = \frac{.557}{.94} = .59$, so H_0 cannot be rejected and inclusion of x_1x_2 as a carrier in the model is not justified.

b No. In the presence of the other four carriers, any particular carrier is relatively unimportant, but this is not equivalent to the statement that all carriers are unimportant.

c $SSE_k = SST(1 - R^2) = 3224.65$, so $f = \left(\frac{5384.18 - 3224.65}{3}\right) / \left(\frac{3224.65}{6}\right) = 1.34$.

Since 1.34 is not $\geq F_{.05,3,6} = 4.76$, H_0 cannot be rejected; the data does not argue for the inclusion of any second-order terms.

45

a $n = 20$, $k = 5$, $n-(k+1) = 14$, so H_0: $\beta_1 = \ldots = \beta_5 = 0$ will be rejected in favor of H_a: at least one among $\beta_1, \ldots, \beta_5$ is $\neq 0$ if $f \geq F_{.01,5,14} = 4.69$. With $f = \left(\frac{.769}{5}\right) / \left(\frac{.231}{14}\right) = 9.32 \geq 4.69$, H_0 is rejected. Wood specific gravity appears to be linearly related to at least one of the five carriers.

b For the full model, adjusted $R^2 = \frac{(19)(.769) - 5}{14} = .687$, while for the reduced model adjusted $R^2 = \frac{(19)(.769) - 4}{15} = .707$.

c From (a), $SSE_k = (.231)(.0196610) = .004542$, and $SSE_1 = (.346)(.0196610) = .006803$, so $f = \left(\frac{.002261}{3}\right) \Big/ \left(\frac{.004542}{14}\right) = 2.32$. Since $F_{.05,3,14} = 3.34$ and 2.32 is not ≥ 3.34, we conclude that $\beta_1 = \beta_2 = \beta_4 = 0$.

Section 13.5

47 a $\ln(Q) = Y = \ln(\alpha) + \beta\ln(a) + \gamma\ln(b) + \ln(\varepsilon) = \beta_0 + \beta_1 x_1 + \beta_2 x_2 + \varepsilon'$ where $x_1 = \ln(a)$, $x_2 = \ln(b)$; $\beta_0 = \ln(\alpha)$, $\beta_1 = \beta$, $\beta_2 = \gamma$, and $\varepsilon' = \ln(\varepsilon)$. Thus we transform to $(y, x_1, x_2) = (\ln(Q), \ln(a), \ln(b))$ (take the natural log of the values of each variable) and do a multiple linear regression. A computer analysis gave $\hat{\beta}_0 = 1.5652$, $\hat{\beta}_1 = .9450$, and $\hat{\beta}_2 = .1815$, For $a = 10$ and $b = .01$, $x_1 = \ln(10) = 2.3026$ and $x_2 = \ln(.01) = -4.6052$, from which $\hat{y} = 2.9053$ and $\hat{Q} = e^{2.9053} = 18.27$.

b Again taking the natural log, $Y = \ln(Q) = \ln(\alpha) + \beta a + \gamma b + \ln(\varepsilon)$, so to fit this model it is necessary to take the natural log of each Q value (and not transform a or b) before using multiple regression analysis.

c We simply exponentiate each endpoint: $(e^{.217}, e^{1.755}) = (1.24, 5.78)$.

49

k	R^2	adj. R^2	$C_k = \frac{SSE_k}{s^2} + 2(k+1) - n$
1	.676	.647	138.2
2	.979	.975	2.7
3	.9819	.976	3.2
4	.9824		4

where $s^2 = 5.9825$.

a Clearly the model with $k = 2$ is recommended on all counts.

b No. Forward selection would let x_4 enter first and would not delete it at the next stage.

51 $k+1 = 3$ so $\frac{2(k+1)}{n} = \frac{6}{12} = .5$, and $h_{ii} < .5$ for every i, so no point is judged to have large influence.

Chapter 13

Supplementary

53 **a** We wish to test H_0: $\beta_1 = \beta_2 = 0$ vs. H_a: either $\beta_1 \neq 0$ or $\beta_2 \neq 0$. The test statistic is $f = \left(\frac{R^2}{k}\right) / \left(\frac{1-R^2}{n-k-1}\right)$, where $k = 2$ for the quadratic model. The rejection region is $f \geq F_{\alpha,k,n-k-1} = F_{.01,2,5}$ (for $\alpha = .01$) = 13.27.

$R^2 = 1 - \frac{.29}{202.88} = .9986$, giving $f = 1783$. No question about it, folks – the quadratic model is useful!

b The relevant hypotheses are H_0: $\beta_2 = 0$ vs. H_a: $\beta_2 \neq 0$. The test statistic value is $t = \frac{\hat{\beta}_2}{s_{\hat{\beta}_2}}$, and H_0 will be rejected at level .001 if either $t \geq 6.869$ or $\leq$ -6.689 (df = $n-3 = 5$). Since $t = \frac{-.00163141}{.00003391} = -48.1 \leq -6.689$, H_0 is rejected. The quadratic predictor should be retained.

c No. R^2 is extremely high for the quadratic model, so the marginal benefit of including the cubic predictor would be essentially nil – and a scatter plot doesn't show the type of curvature associated with a cubic model.

d $t_{.025,5} = 2.571$, and $\hat{\beta}_0 + \hat{\beta}_1(100) + \hat{\beta}_2(100)^2 = 21.36$, so the CI is $21.36 \pm (2.571)(.1141) = 21.36 \pm .29 = (21.07, 21.65)$

e $s^2 = \frac{SSE}{n-3} = \frac{.29}{5} = .058$, so the PI is

$21.36 \pm (2.571)[.058 + (.1141)^2]^{1/2} = 21.36 \pm .69 = (20.67, 22.05)$

55 **a** H_0: $\beta_1 = \beta_2 = 0$ will be rejected in favor of H_a: either $\beta_1 \neq 0$ or $\beta_2 \neq 0$ if $f = \left(\frac{R^2}{k}\right) / \left(\frac{1-R^2}{n-k-1}\right) \geq F_{.01,k,n-k-1} = F_{.01,2,7} = 9.55$. $SST = \frac{\Sigma y_i^2 - (\Sigma y_i)^2}{n} = 264.5$, so $R^2 = 1 - \frac{26.98}{264.5} = .898$ and $f = \left(\frac{.898}{2}\right) / \left(\frac{.102}{7}\right) = 30.8$. Because 30.8 ≥ 9.55, H_0 is rejected at significance level .01 and the quadratic model is judged useful.

b The hypotheses are H_0: $\beta_2 = 0$ vs. H_a: $\beta_2 \neq 0$. The test statistic value is $t = \frac{\hat{\beta}_2}{s_{\hat{\beta}_2}} = \frac{-2.3621}{.3073} = -7.69$ and $t_{.0005,7} = 5.408$, so H_0 is rejected at level .001 and P-value < .001. The quadratic predictor should not be eliminated.

c $x = 1$ here, and $\hat{\mu}_{Y \cdot 1} = \hat{\beta}_0 + \hat{\beta}_1(1) + \hat{\beta}_2(1)^2 = 45.96$. $t_{.025,7} = 1.895$, giving the C.I. $45.96 \pm (1.895)(1.031) = (44.01, 47.91)$.

Chapter 13

57 **a** Estimate $= \hat{\beta_0} + \hat{\beta_1}(15) + \hat{\beta_2}(3.5) = 180 + (1)(15) + (10.5)(3.5) = 231.75$

b $R^2 = 1 - \dfrac{117.4}{1210.30} = .903$

c H_0: $\beta_1 = \beta_2 = 0$ vs. H_a: either β_1 or β_2 (or both) $\neq 0$

$f = \dfrac{.903/2}{.097/9} = 41.9$, which greatly exceeds $F_{.01,2,9}$, so there appears to be a useful linear relationship.

d $s^2 = \dfrac{117.40}{12 - 3} = 13.044$,

$\sqrt{s^2 + (\text{est. st. dev.})^2} = 3.806$, $t_{.025,9} = 2.262$

The P.I. is $229.5 \pm (2.262)(3.806) = (220.9, 238.1)$

59 There are obviously several reasonable choices in each case. In (a), the model with 6 carriers is a defensible choice on all three grounds, as are those with 7 and 8 carriers. The models with 7, 8, or 9 carriers in (b) merit serious consideration. These models merit consideration because R_k^2, MSE_k, and CK meet the variable selection criteria given in Section 13.5.

61 **a** $f = \dfrac{.90/15}{.10/4} = 2.4$. Because $2.4 < 5.86$, H_0: $\beta_1 = \ldots = \beta_{15} = 0$ cannot be rejected. There does not appear to be a useful linear relationship.

b The high R^2 value resulted from saturating the model with predictors. In general, one would be suspicious of a model yielding a high R^2 value when k is large relative to n.

c $\dfrac{R^2/15}{(1 - R^2)/4} \geq 5.86$ iff $\dfrac{R^2}{1 - R^2} \geq 21.975$ iff $R^2 \geq \dfrac{21.975}{22.975} = .9565$

63 **a** The set x_1, x_3, x_4, x_5, x_6, x_8 includes both x_1, x_4, x_5, x_8 and x_1, x_3, x_5, x_6, so $R^2_{1,3,4,5,6,8} \geq \max(R^2_{1,4,5,8}, R^2_{1,3,5,6}) = .723$

b $R^2_{1,4} \leq R^2_{1,4,5,8} = .723$, but it is not necessarily $\leq .689$ since x_1, x_4, is not a subset of x_1, x_3, x_5, x_6.

CHAPTER 14

Section 14.1

1 H_0 will be rejected at level .05 if $\chi^2 \geq \chi^2_{.05,3} = 7.815$. The expected cell counts are $np_{io} = 120p_{io} = 48$, 30, 30 and 12 for $i = 1, 2, 3, 4$, so

$$\chi^2 = \frac{(52-48)^2}{48} + \frac{(38-30)^2}{30} + \frac{(21-30)^2}{30} + \frac{(9-12)^2}{12} = .333 + 2.133 + 27.00 + .750 = 5.916.$$

Since 5.916 is not $\geq$ 7.815, do not reject H_0.

3 We reject H_0 if $\chi^2 \geq \chi^2_{.10,9} = 14.684$.

obs	4	15	23	25	38	31	32	14	10	8
exp	6.67	13.33	20	26.67	33.33	33.33	26.67	20	13.33	6.67
χ^2	1.069	.209	.450	.105	.654	.163	1.065	1.800	.832	.265

$\chi^2 = 1.069 + \ldots + .265 = 6.612$, which is not $\geq$ 14.684, so H_0 is not rejected.

5 H_0: $p_1 = p_2 = p_3 = p_4 = .25$

H_a: at least one proportion $\neq$ .25

$df = 3$

RR: $\chi^2 > 11.344$

Cell	1	2	3	4
Observed	328	334	372	327
Expected	340.25	340.25	340.25	340.25
$\frac{(O-E)^2}{E}$	0.4410	0.1148	2.9627	0.5160

$$\chi^2 = \sum \frac{(O_i - E_i)^2}{E_i} = 4.0345$$

with 3 df, p-value > .10

Fail to reject H_0. The data fails to indicate a seasonal relationship with incidence of violent crime.

Chapter 14

7 **a** Denoting the 5 intervals by $[0, c_1), [c_1, c_2),\ldots,[c_4, \infty)$, we wish c_1 for which $.2 = P(0 \le X \le c_1) = \int_0^{c_1} e^{-x}\,dx = 1 - e^{-c_1}$, so $c_1 = -\ln(.8) = .2231$.

Then $.2 = P(c_1 \le X \le c_2) \Rightarrow .4 = P(0 \le X_1 \le c_2) = 1 - e^{-c_2}$, so $c_2 = -\ln(.6) = .5108$. Similarly, $c_3 = -\ln(.4) = .9163$ and $c_4 = -\ln(.2) = 1.6094$. The resulting intervals are [0, .2231), [.2231, .5108), [.5108, .9163), [.9163, 1.6094), and [1.6094, ∞).

b Each expected cell count is 40(.2) = 8 and the observed cell counts are 6, 8, 10, 7, and 9, so $\chi^2 = \frac{(6-8)^2}{8} + \ldots + \frac{(9-8)^2}{8} = 1.25$. Because $\chi^2_{.10,4} = 7.779$ and 1.25 is not $\ge$ 7.779, even at level .10 H_0 cannot be rejected; the data is quite consistent with the specified exponential distribution.

9 **a** The six intervals must be symmetric about 0, so denote the fourth, fifth, and sixth intervals by $[0, a)$, $[a, b)$, and $[b, \infty)$; a must be such that $\Phi(a) = .6667\left(\frac{1}{2} + \frac{1}{6}\right)$, which from Table A.3 gives $a \approx .43$. Similarly $\Phi(b) = .8333$ implies $b \approx .97$, so the six intervals are (-∞, -.97), [-.97, -.43), [-.43, 0), [0, .43), [.43, .97), and [.97, ∞).

b The six intervals are symmetric about the mean of .5. From (a), the fourth interval should extend from the mean to .43 sd's above the mean, i.e., from .5 to $.5 + .43(.002)$, which gives [.5, .50086). Thus the third interval is $[.5 - .00086, .5) = [.49914, .5)$. Similarly, the upper endpoint of the fifth interval is $.5 + .97(.002) = .50194$, and the lower endpoint of the second interval is $.5 - .00194 = .49806$. The resulting intervals are (-∞, .49806), [.49806, .49914), [.49914, .5), [.5, .50086), [.50086, .50194), and [.50194, ∞).

c Each expected cell count is $45\left(\frac{1}{6}\right) = 7.5$, and the observed counts are 13, 6, 6, 8, 7 and 5, so $\chi^2 = 5.53$. Since $\chi^2_{.05,5} = 11.070$ and 5.533 is not $\ge$ 11.070, H_0 cannot be rejected at level .05 (or at level .10, since $\chi^2_{.10,5} = 9.236$).

Section 14.2

11 According to the stated model, the three cell probabilities are $(1-p)^2$, $2p(1-p)$, and p^2, so we wish the value of p which maximizes $(1-p)^{2n_1}[2p(1-p)]^{n_2}p^{2n_3}$. Proceeding as in example 14.6 gives $\hat{p} = \frac{n_2 + 2n_3}{2n} = \frac{234}{2776} = .0843$. The estimated expected cell counts are then $n(1-\hat{p})^2 = 1163.85$, $n[2\hat{p}(1-\hat{p})] = 214.29$, and $n\hat{p}^2 = 9.86$. This gives $\chi^2 = \frac{(1212-1163.85)^2}{1163.85} + \frac{(118-214.29)^2}{214.29} + \frac{(58-9.86)^2}{9.86} = 280.3$. According to (14.15), H_0 is rejected if $\chi^2 \geq \chi^2_{\alpha,2}$, and since $\chi^2_{.01,2} = 9.210$, H_0 is resoundingly rejected; the stated model is strongly contradicted by the data.

13 The part of the likelihood involving θ is $[(1-\theta)^4]^{n_1} \cdot [\theta(1-\theta)^3]^{n_2}$. $[\theta^2(1-\theta)^2]^{n_3}[\theta^3(1-\theta)]^{n_4}[\theta^4]^{n_5} = \theta^{n_2+2n_3+3n_4+4n_5}(1-\theta)^{4n_1+3n_2+2n_3+n_4} = \theta^{233}(1-\theta)^{367}$, so $\ln(\text{likelihood}) = 233\ln\theta + 367\ln(1-\theta)$. Differentiating and equating to 0 yields $\hat{\theta} = \frac{233}{600} = .3883$, $1-\hat{\theta} = .6117$ [note that the exponent on θ is simply the total # of successes (defectives here) in the $n = 4(150) = 600$ trials]. Substituting this $\hat{\theta}$ into the formula for p_i yields estimated cell probabilities .1400, .3555, .3385, .1433, and .0227. Multiplication by 150 yields the estimated expected cell counts as 21.00, 53.33, 50.78, 21.50, and 3.41. The last estimated expected cell count is less than 5, so we combine the last two categories into a single one ($\geq$ 3 defectives), yielding estimated counts 21.00, 53.33, 50.78, 24.91, observed counts 26, 51, 47, 26, and $\chi^2 = 1.62$. Since $\chi^2_{.10,4-1-1} = \chi^2_{.10,2} = 4.605$ and $1.62 \leq 4.605$, H_0 is not rejected.

15 $\hat{\lambda} = \frac{380}{120} = 3.167$ so $\hat{p}(x) = e^{-3.167}\frac{(3.167)^x}{x!}$.

x	0	1	2	3	4	5	6	≥ 7
$\hat{p}$	.0421	.1334	.2113	.2230	.1766	.1119	.0590	.0427
$n\hat{p}$	5.05	16.00	25.36	26.76	21.19	13.43	7.08	5.12
obs:	24	16	16	18	15	9	6	16

The resulting value of χ^2 is 103.98, and when compared to $\chi^2_{.01,7} = 18.474$, it is obvious that the Poisson model fits very poorly.

17 With $A = 2n_1 + n_4 + n_5$, $B = 2n_2 + n_4 + n_6$, and $C = 2n_3 + n_5 + n_6$, the likelihood is proportional to $\theta_1^A\theta_2^B(1-\theta_1-\theta_2)^C$, where $A+B+C = 2n$. Taking the natural log and equating both $\frac{\partial}{\partial\theta_1}$ and $\frac{\partial}{\partial\theta_2}$ to zero gives $\frac{A}{\theta_1} = \frac{C}{1-\theta_1-\theta_2}$ and $\frac{B}{\theta_2} = \frac{C}{1-\theta_1-\theta_2}$, whence $\theta_2 = \frac{B\theta_1}{A}$. Substituting this into the first equation gives $\theta_1 = \frac{A}{A+B+C}$, and then $\theta_2 = \frac{B}{A+B+C}$. Thus $\hat{\theta}_1 = \frac{2n_1+n_4+n_5}{2n}$, $\hat{\theta}_2 = \frac{2n_2+n_4+n_6}{2n}$, and $(1-\hat{\theta}_1-\hat{\theta}_2) = \frac{2n_3+n_5+n_6}{2n}$. Substituting the observed n_i's yields

$\hat{\theta}_1 = \frac{2(49)+20+53}{400} = .4275$, $\hat{\theta}_2 = \frac{110}{400} = .2750$, and $1-\hat{\theta}_1-\hat{\theta}_2 = .2975$, from which

$\hat{p}_1 = (.4275)^2 = .183$, $\hat{p}_2 = .076$, $\hat{p}_3 = .089$, $\hat{p}_4 = 2(.4275)(.275) = .235$,
$\hat{p}_5 = .254$, $\hat{p}_6 = .164$.

Category	1	2	3	4	5	6
np	36.6	15.2	17.8	47.0	50.8	32.8
Observed	49	26	14	20	53	38

This gives $\chi^2 = 29.1$. With $\chi^2_{.01,6-1-2} = \chi^2_{.01,3} = 11.344$ and $\chi^2_{.01,6-1} = \chi^2_{.01,5} = 15.085$, according to (14.15) H_0 must be rejected since $29.1 \ge 15.085$.

19 MINITAB gives $r = .967$, though the hand-calculated value may be slightly different because when there are ties among the $x_{(i)}$'s, MINITAB uses the same y_i for each $x_{(i)}$ in a group of tied values. $c_{.10} = .9707$ and $c_{.05} = .9639$, so $.05 < P\text{-value} < .10$. At the 5% significance level, one would have to consider population normality plausible.

Section 14.3

21 H_0: TV watching and physical fitness are independent of each other
H_a: The two variables are not independent
$df = (4-1)(2-1) = 3$
with $\alpha = 0.05$, RR: $\chi^2 \ge 7.815$
Computed $\chi^2 = 6.161$
Fail to reject H_0. The data fails to indicate an association between daily TV viewing habits and physical fitness.

23 Let p_{i1} = the probability that a fruit given treatment 1 matures and p_{i2} = the probability that a fruit given treatment 1 aborts. Then H_0: $p_{i1} = p_{i2}$ for i = 1, 2, 3, 4, 5 will be rejected if $\chi^2 \ge \chi^2_{.01,4} = 13.277$.

Chapter 14

Observed matured	Observed aborted		Estimated Expected matured	Estimated Expected aborted	n_i
141	206		110.7	236.3	347
28	69		30.9	66.1	97
25	73		31.3	66.7	98
24	78		32.5	69.5	102
20	82		32.5	69.5	102
			238	508	746

Thus $\chi^2 = \frac{(141-110.7)^2}{110.7} + \dots + \frac{(82-69.5)^2}{69.5} = \mathbf{24.82}$. Since 24.82 is $\geq$ 13.277, H_0 is rejected at level .01.

25 With p_{ij} denoting the probability of a type j response when treatment i is applied, H_0: $p_{1j} = p_{2j} = p_{3j} = p_{4j}$ for j = 1, 2, 3, 4 will be rejected at level .005 if $\chi^2 \geq \chi^2_{.005,9} = \mathbf{23.587}$.

$\hat{E}_{ij}$	1	2	3	4
1	24.1	10.0	21.6	40.4
2	25.8	10.7	23.1	43.3
3	26.1	10.8	23.4	43.8
4	30.1	12.5	27.0	50.5

χ^2 = 27.66 $\geq$ 23.587, so reject H_0 at level .005.

27

$\hat{E}_{ij}$	< .25	.25 - .49	.50 - .74	.75 - .99	≥ 1.0	
P	10.8	23.4	21.4	10.1	14.3	80
A	33.2	71.6	65.6	30.9	43.7	245
	44	95	87	41	58	325

$\chi^2 = 12.37$

Since $12.37 \geq \chi^2_{.05,4} = 9.488$, H_0: $p_{ij} = p_{i.} = p_{.j}$ is rejected in favor of the conclusion that condition and dosage level are dependent.

29 $\chi^2 = \frac{(479-494.4)^2}{494.4} + \frac{(173-151.5)^2}{151.5} + \frac{(119-125.2)^2}{125.2} + \frac{(214-177.0)^2}{177.0} + \frac{(47-54.2)^2}{54.2} + \frac{(15-44.8)^2}{44.8} + \frac{(172-193.6)^2}{193.6} + \frac{(45-59.3)^2}{59.3} + \frac{(85-49.0)^2}{49.0} = 64.65 \geq \chi^2_{.01,4} = 13.277$, so the independence hypothesis is rejected in favor of the conclusion that political views and level of marijuana usage are related.

31 This is a 3 × 3 × 3 situation, so there are 27 cells. Only the total sample size n is fixed in advance of the experiment, so there are 26 freely determined cell counts. We must estimate $p_{..1}, p_{..2}, p_{..3}, p_{.1.}, p_{.2.}, p_{.3.}, p_{1..}, p_{2..}$, and $p_{3..}$, but $\Sigma p_{i..} = \Sigma p_{.j.} = \Sigma p_{..k} = 1$ so only 6 independent parameters are estimated. The rule for df now gives χ^2 df = $26 - 6 = 20$.

33 **a**

Observed			
13	19	28	60
7	11	22	40
20	30	50	100

Estimated	Expected	
12	18	30
8	12	20

$\chi^2 = \frac{(13-12)^2}{12} + \ldots + \frac{(22-20)^2}{20} = .6806$. Because $.6806 < \chi^2_{.10,2} = 4.605$, H_0 is not rejected.

b Each observed count here is ten times what it was in (a), and the same is true of the estimated expected counts, so now $\chi^2 = 6.806 \geq 4.605$, and H_0 is rejected. With the much larger sample size, the departure from what is expected under H_0, the independence hypothesis, is statistically significant – it cannot be explained just by random variation.

c The observed counts are $.13n$, $.19n$, $.28n$, $.07n$, $.11n$, $.22n$, whereas the estimated expected $\frac{(.60n)(.20n)}{n} = .12n, .18n, .30n, .08n, .12n, .20n$, yielding $\chi^2 = .006806n$. H_0 will be rejected at level .10 iff $.006806n \geq 4.605$, i.e. iff $n \geq 676.6$, so the minimum n is 677.

Chapter 14

Supplementary

35 Let p_{i1} = the proportion of fish receiving treatment i (i = 1, 2, 3) who are parasitized. We wish to test H_0: $p_{11} = p_{21} = p_{31}$. With df = **3 − 1 = 2**, H_0 will be rejected at level .01 if $\chi^2 \geq$ **9.210.**

Observed			Estimated	Expected
30	3	33	22.99	10.01
16	8	24	16.72	7.28
16	16	32	22.29	9.71
62	27	89		

This gives χ^2 = **13.1**. Because 13.1 ≥ 9.210, H_0 should be rejected. The proportion of fish that are parasitized does appear to depend on which treatment is used.

37 **a** H_0: The probability of a late-game leader winning is independent of the sport played.
H_a: The two variables are not independent.
Computed χ^2 with 3 *df* = 10.518
RR with α = **0.05:** $\chi^2 \geq$ **7.815** also *P*-value is less than 0.05.
Reject H_0: The two variables appear to be dependent.

b Baseball had fewer than expected late-game leader losses.

39 The estimated expected counts are displayed below, from which χ^2 = 197.70. A glance at the 6 df row of Table A.6 shows that this test statistic value is highly significant – the hypothesis of independence is clearly implausible.

Age	Home	Acute	Chronic	
15 – 54	90.2	372.5	72.3	535
55 – 64	113.6	469.3	91.1	674
65 – 74	142.7	589.0	114.3	846
> 74	157.5	650.3	126.2	934
	504	2081	404	2989

Chapter 14

41 The accompanying table contains both observed and estimated expected counts, the latter in parentheses.

	Age					
Want	127	118	77	61	41	424
	(131.1)	(123.3)	(71.7)	(55.1)	(42.8)	
Don't	23	23	5	2	8	61
	(18.9)	(17.7)	(10.3)	(7.9)	(6.2)	
	150	141	82	63	49	485

This gives χ^2 = **11.60** > $\chi^2_{.05,4}$ = **9.488.** At level .05, the null hypothesis of independence is rejected, though it would not be rejected at level .01 (.01 < *P*-value < .025).

CHAPTER 15

Section 15.1

1 **a** For $a < x < b$, the cumulative distribution function is $F(x) = \frac{x-a}{b-a}$, so $F(\tilde{\mu}) = \frac{\tilde{\mu}-a}{b-a} = .5 \Rightarrow \tilde{\mu} = \frac{a+b}{2}$.

b For $x > 0$, $F(x) = 1-e^{-\lambda x}$, so $F(\tilde{\mu}) = 1-e^{-\lambda\tilde{\mu}} = .5 \Rightarrow \tilde{\mu} = \frac{\ln(2)}{\lambda}$.

3 H_0: $\tilde{\mu} = 6.0$ will be rejected in favor of H_a: $\tilde{\mu} \neq 6.0$ if either $y \geq c$ or $y \leq 10-c$. For $c = 9$, the rejection region is $R = \{0, 1, 9, 10\}$ with $\alpha = .022$, while including 2 and 8 in R results in an α that is too large. With Y = # of X_i's > 6, the observed Y is $y = 2$. Since $y = 2$ is not in the rejection region, H_0 cannot be rejected.

5 **a** With diameter a continuous rv, $p = P(X_{i+6} > X_i) = P(X_{i+6} < X_i)$ and $P(X_{i+6} > X_i) + P(X_{i+6} < X_i) = 1$, which implies that $p = .5$.

b Regarding the determination of whether or not $X_{i+6} > X_i$ as a trial, there are 6 independent trials (since e.g. $X_7 > X_1$ is independent of $X_8 > X_2$) and, from (a), $p = P(S) = .5$, so $Y \sim \text{Bin}(6, .5)$ where Y = number of values of i for which $X_{i+6} > X_i$.

c $P(Y = 6) = b(6;6,.5) = .0156$ and $P(Y = 5) = b(5;6,.5) = .0938$, so with $R = \{5, 6\}$, $\alpha = .0156 + .0938 = .1094 \approx .1$. The observed value of Y is $y = 5$ (the only i for which $x_{i+6} > x_i$ does not hold is $i = 5$, since $x_5 = 5.020 > 5.016 = x_{11}$), which is in R, so H_0 is rejected.

7 With Y = the # of X_i's that exceed 10, $p = P(X_i > 10) = .75$ when H_0 is true (when $\theta = 10$). Thus we wish c such that $\alpha = P(Y \geq c$ when H_0 is true$) = 1 - B(c-1;20,.75) \approx .10$. From the $n = 20$ binomial table $p = .75$ column, $c-1 = 17$ yields $\alpha = 1-.909 = .091 \approx .10$, so $c = 18$ and the rejection region is $R = \{18, 19, 20\}$. Since $y = 18$ (18 of the 20 x_i's exceed 10) is in the rejection region, we reject H_0 at level .1.

Section 15.2

9 We will reject H_0: $\mu = 6$ in favor of H_a: $\mu \neq 6$ if either $s_+ \geq c$ or $\leq \frac{n(n+1)}{2} - c$. With $n = 10$ and $\alpha = .05$, we look for .025 in the $P_0(S_+ \geq c_1)$ column of Table A.9, yielding

$c = 47$ (corresponding to .024), so $\frac{n(n+1)}{2} - c = 8$. H_0 is then rejected if either $s_+ \geq 47$ or ≤ 8. The $(x_i - 6)$'s are -.07, .08, -.14, -.09, .12, -.10, -.05, -.11, -.02, and -.04; since the rank of .08 is 5 and the rank of .12 is 9, $s_+ = 5 + 9 = 14$. Because 14 is neither ≥ 47 nor ≤ 8, H_0 is not rejected.

11 The data is paired and we wish to test H_0: $\mu_D = 0$ vs. H_a: $\mu_D \neq 0$. With $n = 12$ and $\alpha = .05$, $\alpha/2 = .025$, Table A.9 shows that H_0 should be rejected if either $s_+ \geq 64$ or $s_+ \leq 78 - 64 = 14$.

d_i	-.3	2.8	3.9	.6	1.2	-1.1	2.9	1.8	.5	2.3	.9	2.5
rank:	1	10	12	3	6	5	11	7	2	8	4	9

Since $s_+ = 2 + 3 + 4 + 6 + 7 + \ldots + 12 = 72$ and $72 \geq 64$, H_0 is rejected at level .05. In fact, for $\alpha = .01$ (so $\alpha/2 = .005$) the upper-tail critical value is $c = 71$, so even at level .01 H_0 would be rejected.

13 We wish to test H_0: $\mu = 75$ vs. H_a: $\mu > 75$. Since $n = 25$ the large sample approximation must be used, so H_0 will be rejected at level .05 if $z \geq 1.645$. The $(x_i - 75)$'s are -5.5, -3.1, -2.4, -1.9, -1.7, -1.5, -.9, -.8, .3, .5, .7, .8, 1.1, 1.2, 1.2, 1.9, 2.0, 2.9, 3.1, 4.6, 4.7, 5.1, 7.2, 8.7, and 18.7. The ranks of the positive differences are 1, 2, 3, 4.5, 7, 8.5, 8.5, 12.5, 14, 16, 17.5, 19, 20, 21, 23, 24, and 25, so $s_+ = 226.5$ and $\frac{n(n+1)}{4} = 162.5$. Expression (15.2) for σ^2 should be used:

$\tau_1 = \tau_2 = \tau_3 = \tau_4 = 2$, so $\sigma^2 = \frac{25(26)(51)}{24} - \frac{4(1)(2)(3)}{48} = 1381.25 - .50 = 1380.75$

and $\sigma = 37.16$. Thus $z = \frac{226.5 - 162.5}{37.16} = 1.72$. Since 1.72 is ≥ 1.645, H_0 is rejected. $P \approx 1 - \Phi(1.72) = .0427$.

Section 15.3

15 The ordered combined sample is 163(y), 179(y), 213(y), 225(y), 229(x), 245(x), 247(y), 250(x), 286(x), and 299(x), so $w = 5 + 6 + 8 + 9 + 10 = 38$. With $m = n = 5$, Table A.10 gives the upper tail critical value for a level .05 test as 36 (reject H_0 if $W \geq 36$). Since $38 \geq 36$, H_0 is rejected in favor of H_a.

Chapter 15

17 The hypotheses of interest are H_0: $\mu_1 - \mu_2 = 1$ vs. H_a: $\mu_1 - \mu_2 > 1$, where 1 (X) refers to the original process and 2 (Y) to the new process. Thus 1 must be subtracted from each x_i before pooling and ranking.

x-1:	3.5	4.1	4.4	4.7	5.3	5.6	7.5	7.6
Rank:	1	4	5	6	8	10	15	16
***y*:**	3.8	4.0	4.9	5.5	5.7	5.8	6.0	7.0
Rank:	2	3	7	9	11	12	13	14

At level .05, H_0 should be rejected in favor of H_a if $w \geq 84$. Since $w = 65$, H_0 is not rejected.

19

***x*:**	8.2	9.5	9.5	9.7	10.0	14.5	15.2	16.1	17.6	21.5
Rank:	7	9	9	11	12.5	16	17	18	19	20
***y*:**	4.2	5.2	5.8	6.4	7.0	7.3	9.5	10.0	11.5	11.5
Rank:	1	2	3	4	5	6	9	12.5	14.5	14.5

The denominator of z must now be computed according to (15.6). With $\tau_1 = 3$, $\tau_2 = 2$, $\tau_3 = 2$, $\sigma^2 = 175 - .0219[2(3)(4) + 1(2)(3) + 1(2)(3)] = 174.21$, so $z = \dfrac{138.5 - 105}{\sqrt{174.21}} = 2.54$. Because 2.54 is neither ≥ 2.58 nor ≤ -2.58, H_0 cannot be rejected.

Section 15.4

21 With $n = 20$, the sign test that rejects H_0 if either $y \leq 4$ or $y \geq 16$ has level $\alpha = 2(.006) = .012 \approx .01$ (since $B(4;20,.5) = .006$), so with $c = 16$ the sign interval is $(x_{(20-16+1)}, x_{(16)}) = (x_{(5)}, x_{(16)})$. The ordered sample values are 42.5, 43.6, 43.7, 43.9, 44.1, ..., 46.6, 47.0, 47.2, 47.3, and 47.8, so $x_{(5)} = 44.1$, $x_{(16)} = 46.6$, and the CI is (44.1, 46.6).

23 $n = 8$, so from Table A.11 a 95% CI (actually 94.5%) has the form $(\bar{x}_{(36-32+1)}, \bar{x}_{(32)}) = (\bar{x}_{(5)}, \bar{x}_{(32)})$. It is easily verified that the five smallest pairwise averages are $\dfrac{5.0 + 5.0}{2} = 5.00$, $\dfrac{5.0 + 11.8}{2} = 8.40$.

$\frac{5.0+12.2}{2}=8.60$, $\frac{5.0+17.0}{2}=11.00$, and $\frac{5.0+17.3}{2}=11.15$ (the smallest average not involving 5.0 is $\bar{x}_{(6)}=\frac{11.8+11.8}{2}=11.8$), and the five largest averages are 30.6, 26.0, 24.7, 23.95, and 23.80, so the CI is (11.15, 23.80).

25 The ordered d_i's are -13, -12, -11, -7, -6; with $n = 5$ and $\frac{n(n+1)}{2}=15$. Table A.11 shows the 94% CI as (since $c = 1$) $(\bar{d}_{(1)}, \bar{d}_{(15)})$.. The smallest average is clearly $\frac{-13-13}{2}=-13$ while the largest is $\frac{-6-6}{2}=-6$, so the CI is (-13, -6).

27 $m = n = 5$ and from Table A.12, $c = 21$ and the 90% (actually 90.5%) interval is $(d_{ij(5)}, d_{ij(21)})$. The five smallest $x_i - y_j$ differences are -18, -2, 3, 4, 16 while the five largest differences are 136, 123, 120, 107, 87 (construct a table like that of figure 15.7), so the desired interval is (16, 87).

Section 15.5

29 Below we record in parentheses beside each observation the rank of that observation in the combined sample.

1:	5.8 (3)	6.1 (5)	6.4 (6)	6.5 (7)	7.7 (10)	$r_{1.}=31$
2:	7.1 (9)	8.8 (12)	9.9 (14)	10.5 (16)	11.2 (17)	$r_{2.}=68$
3:	5.1 (1)	5.7 (2)	5.9 (4)	6.6 (8)	8.2 (11)	$r_{3.}=26$
4:	9.5 (13)	10.3 (15)	11.7 (18)	12.1 (19)	12.4 (20)	$r_{4.}=85$

H_0 will be rejected at level .10 if $k \ge \chi^2_{.10,3} = 6.251$. The computed value of J is $k=\frac{12}{20(21)}\left[\frac{31^2+68^2+26^2+85^2}{5}\right]-3(21)=14.06$. Since 14.06 is $\ge$ 6.251, reject H_0.

31 H_0: $\mu_1 = \mu_2 = \mu_3$ will be rejected at level .05 if $k \ge \chi^2_{.05,2} = 5.992$. The ranks are 1, 3, 4, 5, 6, 7, 8, 9, 12, 14, for the first sample, 11, 13, 15, 16, 17, 18 for the second, and 2, 10, 19, 20, 21, 22 for the third, so the rank totals are 69, 90, and 94, and $k=\frac{12}{22(23)}\left[\frac{69^2}{10}+\frac{90^2}{6}+\frac{94^2}{6}\right]-3(23)=9.23$. Since 9.23 $\ge$ 5.992, we reject H_0.

Chapter 15

33

	1	2	3	4	5	6	7	8	9	10	r_i	r_i^2
I	1	2	3	3	2	1	1	3	1	2	19	361
H	2	1	1	2	1	2	2	1	2	3	17	289
C	3	3	2	1	3	3	3	2	3	1	24	576
												1226

The computed value of F_r is $\frac{12}{10(3)(4)}(1226) - 3(10)(4) = \mathbf{2.60}$. Since $\chi^2_{.05,2} = 5.992$ and 2.60 is not $\geq$ 5.992, don't reject H_0.

35 Friedman's test is appropriate here. At level .05, H_0 will be rejected if $f_r \geq \chi^2_{.05,3} = \mathbf{7.815}$. It is easily verified that $r_{1.} = 28$, $r_{2.} = 29$, $r_{3.} = 16$, $r_{4.} = 17$, from which the defining formula gives $f_r = 9.62$ and the computing formula gives $f_r = 9.67$. Because $f_r \geq 7.815$, H_0: $\alpha_1 = \alpha_2 = \alpha_3 = \alpha_4 = 0$ is rejected, and we conclude that there are effects due to different years.

37 From Table A.12, $m = n = 5$ implies that $c = 22$ for a confidence level of 95%, so $mn - c + 1 = 25 - 22 + 1 = \mathbf{4}$. Thus the CI extends from the fourth smallest difference to the fourth largest difference. The four smallest differences are -7.1, -6.5, -6.1, -5.9, and the four largest are -3.8, -3.7, -3.4, -3.2, so the CI is (-5.9, -3.8).

39

Sample:	*y*	*x*	*y*	*y*	*x*	*x*	*x*	*y*	*y*
Observations	3.7	4.0	4.1	4.3	4.4	4.8	4.9	5.1	5.6
Rank:	1	3	5	7	9	8	6	4	2

The value of W' for this data is $w' = \mathbf{3 + 6 + 8 + 9 = 26}$. At level .05, the critical value for the upper-tailed test is (Table A.10, $m = 4$, $n = 5$) $c = 27$ ($\alpha = .056$). Since 26 is not $\geq$ 27, H_0 cannot be rejected at level .05.

CHAPTER 16

1 All ten values of the quality statistic are between the two control limits, so no out-of-control signal is generated.

3 P(10 successive points inside the limits) = P(1st inside) × P(2nd inside) × ... P(10th inside) = $(.998)^{10}$ = .9802

P(25 successive points inside the limits) = $(.998)^{25}$ = .9512

$(.998)^{52}$ = .9011 but $(.998)^{53}$ = .8993, so for 53 successive points the probability that at least one will fall outside the control limits when the process is in control is $1 - .8993 = .1007 > .10$.

5 **a** P(point falls outside the limits when $\mu = \mu_0 + .5\sigma$)

$$= 1 - P\left(\mu_0 - \frac{3\sigma}{\sqrt{n}} < \bar{X} < \mu_0 + \frac{3\sigma}{\sqrt{n}} \text{ when } \mu = \mu_0 + .5\sigma\right)$$
$$= 1 - P(-3 - .5\sqrt{n} < Z < 3 - .5\sqrt{n})$$
$$= 1 - P(-4.12 < Z < 1.882) = 1 - .9699 = .0301$$

b
$$1 - P\left(\mu_0 - \frac{3\sigma}{\sqrt{n}} < \bar{X} < \mu_0 + \frac{3\sigma}{\sqrt{n}} \text{ when } \mu = \mu_0 - \sigma\right)$$
$$= 1 - P(-3 + \sqrt{n} < Z < 3 + \sqrt{n})$$
$$= 1 - P(-.76 < Z < 5.24) = .2236$$

c
$$1 - P(-3 - 2\sqrt{n} < Z < 3 - 2\sqrt{n}) = 1 - P(-7.47 < Z < -1.47) = .6808$$

7 $\bar{\bar{x}} = 12.95$ and $\bar{s} = .526$ so with $a_5 = .940$, the control limits are

$$12.95 \pm 3\frac{.526}{.940\sqrt{5}} = 12.95 \pm .75 = 12.20,\ 13.70$$

Again every point $(\bar{x})$ is between these limits, so there is no evidence of an out-of-control process.

9 $\bar{\bar{x}} = \dfrac{2317.07}{24} = 96.54$, $\bar{s} = 1.264$, and $a_6 = .952$, giving the control limits

$$96.54 \pm (3)(1.264)/(.952\sqrt{6}) = 96.54 \pm 1.63 = 94.91,\ 98.17$$

The value of $\bar{x}$ on the 22nd day lies above the *UCL*, so the process appears to be out of control at that time.

Chapter 16

11 **a** $P\left(\mu_0 - \frac{2.81\sigma}{\sqrt{n}} < \bar{X} < \mu_0 + \frac{2.81\sigma}{\sqrt{n}} \text{ when } \mu = \mu_0\right)$
$= P(-2.81 < Z < 2.81) = .995$, so the probability that a point falls outside the limits is .005 and $ARL = \frac{1}{.005} = 200$.

b $p = P$(a point is outside the limits)
$= 1 - P\left(\mu_0 - \frac{2.81\sigma}{\sqrt{n}} < \bar{X} < \mu_0 + \frac{2.81\sigma}{\sqrt{n}} \text{ when } \mu = \mu_0 + \sigma\right)$
$= 1 - P(-2.81 - \sqrt{n} < Z < 2.81 - \sqrt{n})$
$= 1 - P(-4.81 < Z < .81) = 1 - .791 = .209$
Thus $ARL = \frac{1}{.209} = 4.78$

c $1 - .9974 = .0026$ so $ARL = \frac{1}{.0026} = 384.62$ for an in-control process, and when $\mu = \mu_0 + \sigma$, the probability of an out-of-control point is
$1 - P(-3 - 2 < Z < 1) = 1 - P(Z < 1) = .1587$, so $ARL = \frac{1}{.1587} = 6.30$.

13 $\bar{\bar{x}} = 12.95$, $IQR = .4273$, $k_5 = .990$. The control limits are
$12.95 \pm \frac{(3)(.4273)}{.990\sqrt{5}} = 12.37, 13.53$.

15 **a** $\bar{r} = \frac{85.2}{30} = 2.84$, $b_4 = 2.058$, and $c_4 = .880$. Since $n = 4$, $LCL = 0$ and UCL
$= 2.84 + \frac{3(.880)(2.84)}{2.058} = 2.84 + 3.64 = 6.48$

b $\bar{r} = 3.54$, $b_8 = 2.844$, $c_8 = .820$, and the control limits are
$3.54 \pm \frac{3(.820)(3.54)}{2.844} = 3.54 \pm 3.06 = .48, 6.60$

17 $\bar{s} = 1.2642$, $a_6 = .952$, and the control limits are
$1.2642 \pm 3(1.2642)\sqrt{1 - (.952)^2}/.952 = 1.2642 \pm 1.2194 = .045, 2.484$.
The smallest s_i is $s_{20} = .75$ and the largest is $s_{12} = 1.65$, so every s_i is between .045 and 2.434. The process appears to be in control with respect to variability.

Chapter 16

Section 16.4

19 $\bar{p} = \Sigma\frac{\hat{p}_i}{k}$ where $\Sigma\hat{p}_i = \frac{x_1}{n} + \ldots + \frac{x_k}{n} = \frac{x_1 + \ldots + x_k}{n} = \frac{578}{100} = 5.78$. Thus $\bar{p} = \frac{5.78}{25} = .231$

a The control limits are $.231 \pm 3\sqrt{(.231)(.769)/100} = .231 \pm .126 = .105, .357$.

b $\frac{13}{100} = .130$, which is between the limits, but $\frac{39}{100} = .390$, which exceeds the upper control limit and therefore generates an out-of-control signal.

21 $LCL > 0$ when $\bar{p} > 3\sqrt{\bar{p}(1-\bar{p})/n}$, i.e. (after squaring both sides) $50\bar{p}^2 > 3\bar{p}(1-\bar{p})$, i.e. $50\bar{p} > 3(1-\bar{p})$, i.e. $53\bar{p} > 3$, i.e. $\bar{p} > \frac{3}{53} = .0566$.

23 $\Sigma x_i = 102$, $\bar{x} = 4.08$, and $\bar{x} \pm 3\sqrt{\bar{x}} = 4.08 \pm 6.06 \approx (-2.0, 10.1)$
Thus $LCL = 0$ and $UCL = 10.1$. Because no x_i exceeds 10.1, the process is judged to be in control.

25 With $u_i = x_i/g_i$, the u_i's are 3.75, 3.33, 3.75, 2.50, 5.00, 5.00, 12.50, 12.00, 6.67, 3.33, 1.67, 3.75, 6.25, 4.00, 6.00, 12.00, 3.75, 5.00, 8.33 and 1.67 for $i = 1, \ldots, 20$, giving $\bar{u} = 5.5125$.

For $g_i = .6$, $\bar{u} \pm 3\sqrt{\bar{u}/g_i} = 5.5125 \pm 9.0933$, $LCL = 0$, $UCL = 14.6$

$g_i = .8$, $\bar{u} \pm 3\sqrt{\bar{u}/g_i} = 5.5125 \pm 7.857$, $LCL = 0$, $UCL = 13.4$

$g_i = 1.0$, $\bar{u} \pm 3\sqrt{\bar{u}/g_i} = 5.5125 \pm 7.0436$, $LCL = 0$, $UCL = 12.6$

Several u_i's are close to the corresponding UCL's, but none exceed them, so the process is judged to be in control.

Section 16.5

27 $\mu_0 = 16$, $k = \Delta/2 = 0.05$, $h = .20$

$d_i = \max(0, d_{i-1} + (\bar{x}_i - 16.05))$

$e_i = \max(0, e_{i-1} - (\bar{x}_i - 15.95))$

i	$\bar{x}_i - 16.05$	d_i	$\bar{x}_i - 15.95$	e_i
1	-.058	0	.042	0
2	.001	.001	.101	0
3	.016	.017	.116	0
4	-.138	0	-.038	.038
5	-.020	0	.080	0
6	.010	.010	.110	0
7	-.068	0	.032	0
8	-.151	0	-.054	.054
9	-.012	0	.088	0
10	.024	.024	.124	0
11	-.021	.003	.079	0
12	-.115	0	-.015	.015
13	-.018	0	.082	0
14	-.090	0	.010	0
15	.005	.005	.105	0

For no time r is it the case that $d_r > .20$ or that $e_r > .20$, so no out-of-control signals are generated.

29 Connecting 600 on the in-control *ARL* scale to 4 on the out-of-control scale and extending to the k' scale gives $k' = .87$. Thus

$$k' = \frac{\Delta/2}{\sigma/\sqrt{n}} = \frac{.002}{.005/\sqrt{n}}$$

from which $\sqrt{n} = 2.175$, $n = 4.73 = s$. Then connecting .87 on the k' scale to 600 on the out-of-control ARL scale and extending to h' gives $h' = 2.8$, so

$$h = (\sigma\sqrt{n})(2.8) = (.005/\sqrt{5})(2.8) = .00626$$

Chapter 16

Section 16.6

31 For the binomial calculation, $n = 50$ and we wish

$$P(X \le 2) = \binom{50}{0}p^0(1-p)^{50} + \binom{50}{1}p^1(1-p)^{49} + \binom{50}{2}p^2(1-p)^{48}$$
$$= (1-p)^{50} + 50p(1-p)^{49} + 1225p^2(1-p)^{48}$$

when $p = .01, .02, \ldots, .10$. For the hypergeometric calculation,

$$P(X \le 2) = \frac{\binom{M}{0}\binom{500-M}{50}}{\binom{500}{50}} + \frac{\binom{M}{1}\binom{500-M}{49}}{\binom{500}{50}} + \frac{\binom{M}{2}\binom{500-M}{48}}{\binom{500}{50}}$$

to be calculated for $M = 5, 10, 15, \ldots, 50$. The resulting probabilities appear in the answer section in the text.

33

$$P(X \le 2) = \binom{100}{0}p^0(1-p)^{100} + \binom{100}{1}p^1(1-p)^{99} + \binom{100}{2}p^2(1-p)^{98}$$

p	.01	.02	.03	.04	.05	.06	.07	.08	.09	.10
$P(X \le 2)$	.9206	.6767	.4198	.2321	.1183	.0566	.0258	.0113	.0048	.0019

For values of p quite close to 0, the probability of lot acceptance using this plan is larger than that for the previous plan, whereas for larger p this plan is less likely to result in an "accept the lot" decision (the dividing point between "close to zero" and "larger p" is someplace between .01 and .02). In this sense, the current plan is better.

35 P(accepting the lot) = $P(X_1 = 0 \text{ or } 1) + P(X_1 = 2, X_2 = 0, 1, 2, \text{ or } 3) + P(X_1 = 3, X_2 = 0, 1, \text{ or } 2) = P(X_1 = 0 \text{ or } 1) + P(X_1 = 2) \bullet P(X_2 = 0, 1, 2, \text{ or } 3) + P(X_1 = 3) \bullet P(X_2 = 0, 1, \text{ or } 2)$

$p = .01$: $= .9106 + (.0756)(.9984) + (.0122)(.9862) = .9981$

$p = .05$: $= .2794 + (.2611)(.7604) + (.2199)(.5405) = .5968$

$p = .10$: $= .0338 + (.0779)(.2503) + (.1386)(.1117) = .0688$

37 **a** $AOQ = pP(A) = p[(1-p)^{50} + 50p(1-p)^{49} + 1225p^2(1-p)^{48}]$

p	.01	.02	.03	.04	.05	.06	.07	.08	.09	.10
APQ	.010	.018	.024	.027	.027	.025	.022	.018	.014	.011

b $p = .0447$, $AOQL = .0447P(A) = .0274$

c $ATI = 50P(A) + 2000(1 - P(A))$

p	.01	.02	.03	.04	.05	.06	.07	.08	.09	.10
ATI	77.3	202.1	418.6	679.9	945.1	1188.8	1393.6	1559.3	1686.1	1781.6

Supplementary Exercises

39 $n = 6$, $k = 26$, $\Sigma\bar{x}_i = 10{,}980$, $\bar{\bar{x}} = 422.31$, $\Sigma s_i = 402$, $\bar{s} = 15.4615$, $\Sigma r_i = 1074$, $\bar{r} = 41.3077$

S chart: $15.4615 \pm \dfrac{3(15.4615)\sqrt{1-(.952)^2}}{.952}$

$= 15.4615 \pm 14.9141 \approx .55,\ 30.37$

R chart: $41.31 \pm \dfrac{3(.848)(41.31)}{2.536} = 41.31 \pm 41.44$

so $LCL = 0$, $UCL = 82.75$

$\bar{x}$ chart based on $\bar{s}$: $422.31 \pm \dfrac{3(15.4615)}{.952\sqrt{6}} = 402.42,\ 442.20$

$\bar{x}$ chart based on $\bar{r}$: $422.31 \pm \dfrac{3(41.3077)}{2.536\sqrt{6}} = 402.36,\ 442.26$

41

i:	1	2	3	4	5	6
$\bar{x}_i$:	50.83	50.10	50.30	50.23	50.33	51.20
s_i:	1.172	.854	1.136	1.097	.666	.854
r_i:	2.2	1.7	2.1	2.1	1.3	1.7

i:	7	8	9	10	11
$\bar{x}_i$:	50.17	50.70	49.93	49.97	50.13
s_i:	.416	.964	1.159	.473	.698
r_i:	.8	1.8	2.1	.9	.9

i:	12	13	14	15	16	17
$\bar{x}_i$:	49.33	50.23	50.33	49.30	49.90	50.40
s_i:	.833	.839	.404	.265	.854	.781
r_i:	1.6	1.5	.8	.5	1.7	1.4

i:	18	19	20	21	22
$\bar{x}_i$:	49.37	49.87	50.00	50.80	50.43
s_i:	.902	.643	.794	2.931	.971
r_i:	1.8	1.2	1.5	5.6	1.9

$\Sigma s_i = 19.706$, $\bar{s} = .8957$, $\Sigma \bar{x}_i = 1103.85$, $\bar{\bar{x}} = 50.175$
$a_3 = .886$, from which an s chart has $LCL = 0$ and $UCL =$

$.8957 + 3(.8957)\sqrt{1-(.886)^2}/.886 = 2.3020$, $s_{21} = 2.931 > UCL$. Since an assignable cause is assumed to have been identified we eliminate the 21st group. Then

$\Sigma s_i = 16.775$, $\bar{s} = .7998$, $\bar{\bar{x}} = 50.145$, The resulting UCL for an s chart is 2.0529, and $s_i < 2.0529$ for every remaining i. The $\bar{x}$ chart based on $\bar{s}$ has limits

$50.145 \pm 3(.7988)/(.886\sqrt{3}) = 48.58, 51.71$. All $\bar{x}_i$ values are between these limits.

43 $\Sigma n_i = 4(16) + (3)(4) = 76$, $\Sigma n_i\bar{x}_i = 32{,}729.4$, $\bar{\bar{x}} = 430.65$

$$s^2 = \frac{\Sigma(n_i - 1)s_i^2}{\Sigma(n_i - 1)} = \frac{27{,}380.16 + 5661.4}{76 - 20} = 590.0279,\ s = 24.2905$$

For variation: when $n = 3$,

$$UCL = 24.2905 + \frac{3(24.2905)\sqrt{1-(.886)^2}}{.886} = 24.29 + 38.14 = 62.43$$

when $n = 4$,

$$UCL = 24.2905 + \frac{3(24.2905)\sqrt{1-(.921)^2}}{.921} = 24.29 + 30.82 = 55.11$$

For location: when $n = 3$, $430.65 \pm 47.49 = 383.16, 478.14$
$n = 4$, $430.65 \pm 39.56 = 391.09, 470.21$
All values are well within the corresponding control limits.